Engineering Economics

Engineering Economics

Robert L. Mitchell
Senior Lecturer
The University
Zimbabwe–Rhodesia

A Wiley–Interscience Publication

JOHN WILEY & SONS

Chichester · New York · Brisbane · Toronto

British Library Cataloguing in Publication Data

Mitchell, Robert L.
  Engineering economics.
  1. Economics 2. Engineering economy
  I. Title
  330′.02′462   HB171   79-40647

  ISBN 0 471 27640 5 cloth
  ISBN 0 471 27619 7 paper

Photosetting by Thomson Press (India) Limited, New Delhi
and printed by The Pitman Press, Bath.

Contents

Preface

When the author read for his engineering degree, no costing, let alone economics, was taught in the syllabus. Later in life, when a chief engineer with a government ministry, he came to realize that he was unable to justify his designs on economic grounds, and then read for a further degree in economics.

In recent years many engineering faculties of universities have introduced short courses on economics within their engineering degree syllabuses, but their courses are often mounted by economists who fail to speak the engineer's language, and to understand his interests.

The author is presently lecturing in a Faculty of Engineering where he is responsible for teaching engineering economics to third-year students of all disciplines. This text is based on thirty hours of lectures, and, being a mixture of economics and engineering, should be of value both to engineering students, and to practising engineers, unable to study economics in depth.

The author maintains that, whilst a course of microeconomics may be of value to technologists, the engineer needs to master the techniques of cost appraisals and cost–benefit analysis. To do so he must have a working knowledge of the basic principles of both micro and macro economics, of money, of the business cycle, of balance-of-payments, and of taxation; in the interests of brevity the author has sacrificed depth for width in his treatment, and has omitted Keynesian and post-Keynesian economic theory. Examples given are based solely on problems of engineering disciplines.

It is hoped that the work can be easily read and mastered so that the interested student will be able to study further works in any specialist economic field without having to read any other basic text. He suggests that this work is complementary to *An Introduction to Engineering Economics* published by the Institution of Civil Engineers, London, which, in spite of its title, is a reference to cost appraisals rather than to economic theory.

1980   R. L. MITCHELL

General Introduction

THE NEED FOR ECONOMICS

Society operates on political and managerial decisions made, directly or indirectly, on the advice of engineers and economists. In fact, with the complexities of modern society, many other disciplines are involved: ecology, geology, planning, and social science, to name but a few, and it behoves each profession to be able to speak the language of others. More than that, because of the breadth of the engineer's training in applied science, he often finds himself as the coordinator of other disciplines; it should be simpler for him to grasp the basic principles of economics than for the economist to grasp those of engineering.

Moreover the engineer cannot work in an economic vacuum. Even forgetting the old dictum that an engineer can do for one dollar what anyone else can do for two, it is patently obvious that no design can be valid unless it is the most economic design of several alternatives. I guardedly did not say 'cheapest' — there is a world of difference between the *cheapest* and the most *economic* engineering solution — and costing, or cost accountancy, a monetary or financial discipline, must be regarded as a means to an end, not an end in itself.

Let us analyse this statement. Assume we are given two machines of equal output; one costs twice as much as the other but may last say twenty years instead of five. The cheapest is obvious, but which will be the more economic when maintenance, fuel consumption, obsolescence, and the mortgage or borrowing rate is considered? If one is locally made, and one is imported, and there is a shortage of foreign currency due to the balance-of-payments situation, the choice becomes even more complicated. An understanding of the basic laws of economics becomes a prerequisite in analysing such a problem.

Shakespeare's Caesar said 'The fault, dear Brutus, lies not in our stars, but in ourselves, that we are underlings.' The engineer who has no understanding of the laws of economics is likely to remain an underling — he will find employment as a technologist, but not as an engineer in the broad sense; he may well be a contented employee, but will not make a consultant, neither will he become a head of a government department or a company director. Those few who do so without this knowledge are likely to find, according to the Peter principle*, that each man will rise to his own level of incompetence!

THE MEANING OF ECONOMICS

One of many acceptable definitions suggested by Samuelson†, is very apposite to our purpose. This is 'Economics is the study of how men choose to use scarce or limited productive resources (land, labour, capital goods such as machinery, and technical knowledge) to produce various commodities (such as wheat, overcoats, roads, concerts, and yachts) and to distribute them to various members of society for their consumption.'

Engineering economics may then be defined as the study of how engineers choose to optimize their designs and construction methods to produce designs or objects which will optimize their efficiency and hence the satisfaction of their clients.

Economics is conventionally and conveniently split into two main streams: *micro economics*, the study of economic laws on a small scale as affecting a firm; and *macro economics*, the study on the national and international scale, as affectng the wealth of society.

There are also many subdivisions: of *money*, and of *social* or *welfare economics*; and these are dealt with more briefly although they have many applications, particularly in civil engineering.

Some knowledge of all these divisions of economics is necessary before the engineer can fully grasp the principles of *project appraisal* and *cost–benefit analysis* without which little engineering work can be justified.

THE APPLICATION OF ENGINEERING ECONOMICS

Whilst all engineers agree on the need to study economics, there is often confusion between economics and costing. However, once this confusion has been dispelled, there is little doubt as to the relative importance of micro and macro economics as a background to cost appraisal and cost–benefit analysis (CBA).

If one takes a narrow view of the engineer as a factory executive, or perhaps as a civil engineering contractor, then he needs a thorough understanding of micro economics, i.e. the theory of production, supply and demand, and also inflation and the business cycle. However, many engineers may become senior civil servants and consultants to governments. In this case they will need an understanding of macro economics, i.e. fiscal and economic controls, the national product, welfare economics, taxation, and of balance-of-payments problems. All need a thorough understanding of project cost analysis.

In so short a book it is impossible to cover the whole field in depth. This work merely scratches the surface of highlights and fundamentals of the science of economics in the hope that the reader will gain enough knowledge to understand

* Peter, L. J., and Hull, R., *The Peter Principle*. London, Souvenir Press (1969).
† Samuelson, P. A., *Economics, an Introductory Analysis*. New York, McGraw-Hill (1976).

his colleague, the economist, and, if his interest is awakened, he will then be able to delve deeper into specialist works at leisure.

Finally it is necessary to stress to engineers, trained as physical scientists, that economics is a social science, even if it is possibly the social science which most approaches a physical discipline. Since the mathematics of economics is a quantification and projection of observed data, at best it can only give answers as accurate as the input, which, by its very nature, cannot be exact. Indeed, as solutions in engineering are seldom accurate to within 1%, it is postulated that those in economics may not be accurate tho within 10%, certainly where cost—benefit analysis, which deals with the future, is involved; and this aspect is stressed throughout the work. Solutions to many problems may be reached by differential analysis, but have been avoided where possible as they infer unjustified accuracy.

Chapter 1

Introduction to Micro Economics

1.1 HISTORICAL ORIGINS

Modern economics has developed from the Classical School, the 'father' of which was Adam Smith, whose classic *The Wealth of Nations* was published in 1776.

Two important concepts derived from this. The first was the theory of specialization, which is most pertinent to engineers, as it underlies the concept of mass production. On examining the manufacture of pins Smith found that one man could make only a few dozen each day. However, when a group of men got together, one cutting wire, one grinding it, one making heads etc., production could be increased a thousand-fold. This concept of specialization has, of course, subsequently led to mechanization.

The second concept is no less important. Smith conceived the principle of an 'Invisible Hand', whereby every individual, selfishly pursuing his self-interest, was guided by an invisible hand, which resulted in the achievement of the best 'good' for all. It follows that any interference of this 'invisible hand' or of 'free competition' by government would upset the balance and reduce the total common good. Smith did not, of course, appreciate the full implications of this concept, which led to that of 'Perfect Competition'.

1.2 PERFECT COMPETITION

This is defined as the case where no individual producer (or consumer), be he engineer, farmer or labourer, has any personal influence on the market price. It is this state of affairs which underlies basic micro-economic theory, and theories of imperfect competition, where there is government interference in the market, or where one producer or consumer is large enough to affect the market price, are developed from it. Perfect competition is thus assumed hereafter, unless there is specific reference to the contrary.

1.3 COSTS

Before we can undertake economic analysis, we must have a knowledge of costs. It is convenient to break down costs into *fixed costs*, *variable costs*, and *total costs*, before deriving *average costs*.

Fixed Costs

These are inescapable costs, which are unavoidable and do not change with use. For example, if we wish to produce cars, we must have a factory building; this constitutes a fixed cost. This cost must be allocated, that is passed on to the sales price of the cars, as a fixed cost. If our factory costs $N whether we use it to produce one car or a thousand the fixed cost is still $N.

Thus it can be seen that the fixed cost *per car produced* will reduce with increasing production, even though the fixed cost itself remains unchanged.

Variable Costs

These are costs which increase with output (production). For example if our labour costs for producing a car are $M, they are likely to be about $2M for producing two cars, $3M for three cars, etc.

Total Costs

These are the total costs (of production) and are our main interest. Total cost is the sum of fixed and variable costs, and will increase with production.

Average Costs

Average costs are the total costs, calculated for any level of production, divided by the number of units produced. Initially they will decrease with increasing production, but will eventually increase with it as variable costs can increase out of proportion to additional output, as will be shown later.

1.4 COSTING

We are all concerned with transportation costs, and, in particular, with the costs of running a car, so let us consider this case. We are aware that the more we use our car the more it costs us in petrol, but, within practical limits, the less it costs us per kilometre travelled.

What are these practical limits? We know that if we run a car, not for 1 km, but for 10 000, the average cost per kilometre will reduce; it is reasonable too that it will be even less if we run it for 100 000 km. But suppose we run it for 1 000 000 km? Then we know that the maintenance bills will become excessive, for, with increasing use, we are liable to have to face large repair bills for replacing bearings etc., and we know instinctively that it will be cheaper to scrap the car and buy a new (or newer) one, as the cost per *extra* kilometre will be more on an old than on a new car. Eventually, too, the cost per *average* kilometre will be higher on the old car.

In economic terms the *marginal cost* of running each additional kilometre beyond, say, 100 000, will be increasing and the *Law of Diminishing Returns* will apply. These terms are discussed later.

We have to break down our costs into fixed and variable costs, the sum of which is total cost. The problem now is which way to budget for the individual costs. Petrol and oil are variable costs, whilst tax, insurance etc. are fixed costs, in that they have to be paid whether the car is used or not. But what of the cost of the car itself? The car depreciates in value whether we use it or not — thus it should be regarded as a fixed cost. However, owing to obsolescence, it will depreciate, but somewhat less if it is used less, so theoretically some of its depreciation could be budgeted as a variable cost. It is, however, a conventional simplification to consider its depreciation as a fixed cost, and this we will do. Similarly, with inflation it can be argued that a car may not even depreciate — however, this is a fallacious argument, as it is the purchasing power of money which is changing, not the true value of the car.

Example 1.1   Simple costing of running a car

Let us assume that a car costs us $4000, and after running it, we scrap it for nil value. Let us also assume that fuel, tyres, maintenance etc. will cost us 3 cents per kilometre.

We can now calculate Table 1.1, and plot Figure 1.1. As can be seen, the longer that the car can be run until it has no scrap value, the lower will be *average costs*, *AC*, though *total costs*, *TC*, increase. The constant nature of *fixed costs*, *FC*, and the increasing nature of *variable costs*, *VC*, and of the summation, *total costs*, *TC*, can readily be seen. Of more significance is the nature of *average costs*, *AC*, which drop from infinity at zero use to 13 cents per kilometre at 40 000 km use and to 7 cents per kilometre 100 000 km use.

1.5   FURTHER COMPLICATIONS TO COST ANALYSIS

In the real world we have the option at any time of selling our car and buying a new one. In Example 1.1, ignoring interest on capital, we need merely to deduct

Table 1.1   Costs of running a car

Distance travelled	Fixed costs	Variable costs	Total costs $ TC	Average costs cents/km AC
km × 10³	$ FC	$ VC	Col 2 + Col. 3	$\frac{\text{Col. 4}}{\text{Col. 1}}$
Col. 1	Col. 2	Col. 3	Col. 4	Col. 5
0	4000	0	4000	∞
20	4000	600	4600	23.0
40	4000	1200	5200	13.0
60	4000	1800	5800	9.7
80	4000	2400	6400	8.0
100	4000	3000	7000	7.0

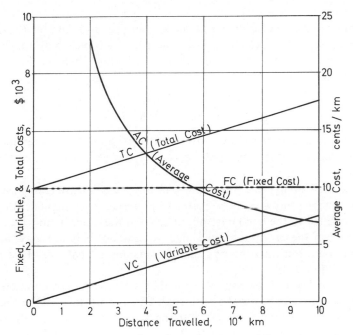

Figure 1.1   Simple costing of running a car. *AC*, average cost;
TC, total cost; *FC*, fixed cost; *VC*, variable cost

the sales price (obtained after any given amount of use) from the purchase price; this will reduce *FC*, *TC*, and *AC*, but of course, we cannot extend the analysis to reduce *AC* beyond the value calculated from the distance travelled at the time of sale.

For example, if we sold after 40 000 km for $2000 (and bought a new car for $4000), our *AC* over the first 40 000 would drop from 13 to 8 cents per kilometre, but it would rise again thereafter dropping again to 8 cents if we resold again for $2000 after a further 40 000 km. It can be seen that, on this basis, it would never drop to 7 cents per kilometre — we would be paying more for the privilege of driving in newer cars.

It might be asked why we should not continue running the old car. The answer is that after a large usage our repair bills would not be constant, but would start escalating. In other words, our variable costs would increase out of proportion to usage — the *VC* and *TC* curves would become concave upwards and eventually so would the *AC* curve.

In this analysis, we have also ignored expenses of taxation, insurance, and garaging. In our imperfect world these are seldom based on distance travelled (the partial exceptions being in those countries wise enough to incorporate vehicle tax with fuel tax which means that tax increases with usage on a pay-as-you-drive, not on a time, basis).

Such costs are inescapable, being time, but not use, dependent; they should be regarded as fixed costs and added to the cost of the car. From this it follows that the commercial driver who does 100 000 km per year will raise his *FC each year* the same amount as will the housewife who perhaps travels only 10 000 km per year, and whose average costs are thus raised more than his by such *time-dependent* fixed costs.

1.6 INTEREST RATES

The major simplification we have introduced is to ignore the market interest rate in our calculation — the *opportunity cost of capital* as it is sometimes called by economists.

If we have $4000 in cash and spend it to buy a car, we are losing the interest which we would otherwise have accrued by investing the capital; this interest should be added to the fixed cost of our car. On the other hand, if we have no capital and buy the car on hire purchase, we have not only to repay the finance house our loan, but also interest on the outstanding balance. This will also increase *FC* and *TC*.

It is also possible that we will pay something as a cash deposit and borrow the rest, repaying our loan before the car is sold. The combinations of this are endless, of course, so we will confine ourselves to the case where we borrow the total sum and repay it over the life of the car. Such a loan is analogous to the more familiar house mortgage.

It is worth mentioning that in the life of a car, if we will need to spend $Y on running costs and the like, then we will need less than $Y available at the time of purchase, as this smaller sum would earn interest. However, this complication is not considered at this stage, as it involves an understanding of *present day values*, discussed in Chapter 17 and following pages.

1.7 MORTGAGES

These are defined as the redemption (repayment) of a loan by *equal* payments, at a given interest rate, over an agreed period. The amount that we pay back will exceed the loan by the interest payable. If we borrow, say, $1000 at 10% interest at an annual repayment of $120 per annum, after the first year we will owe $1000 \times 1.1 = 1100 and, after the first repayment we will owe $1100 - $120 = 980.

House mortgages and hire-purchase loans — essentially the same thing — are normally repayable in equal *monthly* instalments. Suppose, then, we repaid $10 a month, not $120 a year. After one month we will then owe almost (but not quite)* $1000(1 + \frac{10}{1200}) = 1008.3. The next day having repaid $10, we will then only owe $1008.3 - $10 = 998.3 on which we will accumulate an interest debt. It follows that we will repay our mortgage debt more quickly with monthly than with yearly repayments. Note that our first monthly repayment of $10 represents

---

* 10/12% per month interest is not quite the same as 10% per annum, since interest paid each month earns interest.

$1.7 capital repayment and $8.3 interest, whilst it can be calculated that our last repayment of $10 will comprise $9.92 capital and 8 cents interest.

However, in the real world for taxation purposes we have to account yearly, and engineering works are normally long term, so, for convenience we balance our books and cost loans on a yearly basis.

Digression. Consider a mortgage as a redemption of a loan by equal annual payments over n years at an interest rate r on the outstanding balance.

Let a sum P be repaid after each year and the loan, initially $1, in reduced each year by the amount P less the amount of interest accumulating during the year.

Thus the outstanding debt:

$$\text{after year } 1 = 1 + r - P$$
$$\text{after year } 2 = (1 + r - P)(1 + r) - P$$
$$\vdots$$
$$\text{after year } n \doteq (1 + r)^n - P[(1 + r)^{n-1} + (1 + r)^{n-2} + \cdots + 1]$$

which, by definition, is zero. Therefore

$$P = \frac{r}{1 - (1 + r)^{-n}}$$

This calculation is tedious, and can be solved by using Table C. For example, if we borrow $4000 at 15% interest rate over ten years,

$$P = \frac{4000}{5.0188} = \$797 \text{ p.a.}$$

Thus, after one year we owe $(4000 \times 1.15) - \$797 = \$4600 - \$797 = \$3803$; of the $797, $600 is interest and $197 is repayment. After the second year the interest is $3803 \times 0.15 = \$570.5$ so the loan is reduced by $226.5, etc.

1.8 COSTS OF RUNNING A CAR WHERE INTEREST IS INVOLVED

Whether we borrow capital and repay it with interest (a mortgage), or whether we use our own capital, the procedure for costing is the same, as the capital for purchase would not have been left idle; we would withdraw it from a savings account, or from other investment, where it would have been earning interest.

Example 1.2 More complex costing of running a car

Assume the data of Example 1.1, but that the car life is 10 years, with an interest rate of 15% and equal annual repayments over this period. From Table C we can calculate that the annual repayment is

$$\$\frac{4000}{5.0188} = \$797 \text{ p.a.}$$

Thus FC will rise from $4000 to $7970. We can now complete a table, similar to Table 1.1, but with the higher FC, and deduce Figure 1.2. The VC is unaltered but the FC rises to $7970 and the AC rises considerably. It should be noted that since the FC remains above VC (they would become equal at some 265 000 km

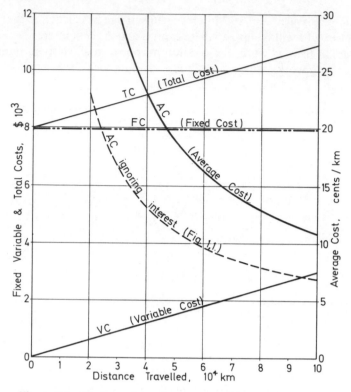

Figure 1.2 Costing of running a car on hire purchase

which exceeds the life of the average car), the fixed costs account for the major part of both total and average costs of car operation, particularly when tax, insurance, and garaging costs are added to them.

In the case where other annual fixed costs must be considered, such as tax and insurance as discussed in Section 1.5, although these are time, not use based, and are still fixed costs, they are not payable when the car is purchased, but must be paid during the use of the car, like variable costs. Thus, in this sense, they are variable and funds for them would not normally be borrowed but paid out of income as are petrol and tyres, so they would not be amortized, or thus subject to interest adjustment.

PROBLEMS

Problem P1.1

Determine the average cost of running a car, costing $4000, for 10 years at 10 000 km per year, with fuel, tyres, and maintenance costing 3 cents per km, and garaging, tax, and insurance costing $250 p.a. It is assumed that the car depreciates 1% every 1000 km (i.e. life is 100 000 km), that the interest rate is 10%, and, for simplification, that all payments are made at the year end.

Problem P1.2

Calculate the cost per km of the car in Problem P1.1 if it is driven at 20 000 km per year, assuming the same life of 100 000 km.

Problem P1.3

The driver in Problem P1.2 decides to travel on public transport for 10 000 km each year at a cost of 4 cents per km. The problem is to calculate his annual savings by so doing.

Problem P1.4

The driver in Problem P1.1 wishes to run a better car, costing $8000, which depreciates less, at 0.5% per 1000 km. He plans to sell this after 10 years, and 100 000 km use. Determine his new average costs.

Chapter 2

The Theory of Production

2.1 FACTORS OF PRODUCTION

In the language of economics, articles produced, sold, or exchanged are known as goods. A good may be tangible or intangible. In order for a good to be produced, four factors (of production) are necessary: labour, land, capital, and enterprise.

Labour includes all labour, from the cleaner to the paid manager. It is obvious that without human labour and skill nothing can be produced.

Land includes not only land (including the location of the land) but buildings — the physical wherewithal of a place where production can be carried out. In the financial, as opposed to the economic sense, buildings can be classified as capital, and depreciated, like machinery for taxation purposes, but this does not alter the principles of economic analysis.

Capital. This is the money or capital available, but in the long term it includes machinery, tools, and materials since, with money, such equipment and materials may be acquired.

Enterprise. This fourth factor is the least tangible of all. This term usually means risk and initiative, whether by a single entrepreneur or by a state. Capital, land, and labour must be organized and given a green light to proceed after the market has been assessed. Perhaps a good example would be the captain and the owners of a tramp steamer, which diverts whilst at sea to pick up a cargo here or there. They are both entrepreneurs — both are involved in initiative and risk. The captain, in so far as he draws a salary, is *also* counted as labour.

Today engineers and economists have become energy conscious, and it can be argued that there is a fifth factor of production — energy. Moreover, being subject to the laws of entropy, this is irreversible. There appears to be no reason, however, why traditional theory needs to be discarded if energy is listed with materials under the definition of capital, as long as its irreversible nature is not overlooked.

2.2 LAWS OF PRODUCTION

Having established the concepts of fixed, variable, total, and average costs, and having analysed the factors necessary for production, two concepts must be considered: the *law of diminishing returns* and *increasing returns to scale*.

The Law of Diminishing Returns

This law states that whilst an increase in input of one factor of production may cause an increase in output, eventually a point will be reached beyond which increasing units of input will cause progressively less increases in output.

Let us postulate a foundry business, manufacturing a standard casting, with spare space capacity for castings, and see what is likely to happen to our production if we increase one factor of production, say labour. In this case, enterprise and land are unchanged.

If one man can produce 10 castings a day, we might expect a second man to

Table 2.1   Illustration of law of diminishing returns

Number of Men	Total daily output	Average daily output per man	Increase in daily output from one extra man*
Col. 1	Col. 2	Col. 3	Col. 4
1	10	10	
			10
2	20	10	
			8
3	28	9.3	
			7
4	35	8.7	
			6
5	41	8.2	
			5
6	46	7.7	
			4
7	50	7.1	
			4
8	54	6.7	
			3
9	57	6.3	
			2
10	59	5.9	
			1
11	60	5.5	
			0
12	60	5.0	
			− 1
13	59	4.5	

*Column 4, the increase in output resulting from the employment of one additional man is known as the *marginal output*. It is important to plot marginal data between lines of equal input.

increase the output to 20, but a third to only 28, a fourth to only 35 etc. Indeed we might expect an output of 60 from 12 men, but only 59 from 13, as floor space has become so crowded that the extra men get in each others' way and hinder efficient production. We can tabulate this in Table 2.1.

In this foundry the law of diminishing returns is applying, since, when we have more than two men, increasing units of the labour force will give rise to progressively less increase in output.

A good example of the workings of this law is often seen on civil engineering sites. Here the space available for concreting operations is often to restricted that if too many labourers are hired to transport the concrete, queues will form and reduce production.

Increasing Returns to Scale

Whilst the law of diminishing returns is a fundamental law of production, it may be hard to accept in the common knowledge that mass production usually reduces cost. However, the law presupposes that not all of the four factors are increased simultaneously; if they are, then theoretically the production could increase *pro rata*.

However, in practice production is often likely to increase at a faster rate than the increase of factors of production—this is called *increasing returns to scale*. This is due to technological factors and the possibility of specialization. Consider our foundry for example. If one man could specialize in heating and pouring metal to meet the needs of say five others, who would concentrate on pattern making, we could achieve increasing returns to scale, the economy of mass production. Six men might then have an output of say seventy castings a day, against the ten casting for one man, or forty six for the six men without specialization.

However, if the firm increased in size, say to eight or nine men, the law of diminishing returns would again apply. We would then have to increase many, even all, the other factors of production. An increase of one pattern maker from five to six would obviously result in no further *pro rata* increase in output, as the metal pourer would become the bottleneck; conversely, hiring one extra metal pourer would also not increase output which would then be restricted by the five pattern makers.

An industry with an output large enough to justify a Research and Development section can streamline its production and can thus also achieve economies of scale.

It can thus be concluded that whilst an increase in one or more factors of production can result in increased output (and hence in lower unit costs) sooner or later the law of diminishing returns will apply and triumph.

2.3   THE SHORT-RUN PRODUCTION FUNCTION

We are now in a position to examine the production of a single firm in the *short run*, i.e. in a period where, say, major factory extensions are not possible, when only limited adjustments are possible to meet changing output needs — as opposed to the *long run* when many, or all, factors of production may be adjusted to meet changing demand.

Production consists of transforming input into outputs and can be written:

$$Q = f(x)$$

where Q is the output and x the input, or more explicitly:

$$Q = f(\bar{x})$$

where $\bar{x}$ is the summation of x_1, x_2, etc., since there may be several inputs.

Let us now consider the output–input relationship (i.e. the production function) of a capacitor-start single-phase electric motor. (This is of a small type such as is used in teaching laboratories.)

Table 2.2 Performance of a laboratory motor

Input,	x,W	112	144	180	220	270	296	324	360	400	452
Output,	Q,W	30.9	61.3	91.1	120	148.5	162	175	187	195	183

The test run produced the data shown in Table 2.2, which are plotted in Figure 2.1(a). In economic analysis, for convenience, we should work in uniformly increasing units of input, so these figures are 'graphed' and recalculated in input increments of 50, as shown in Table 2.3. (It may be noted that the motor stalls before reaching an input of 500 W but for convenience the curve is extrapolated to 500 W input.) Column 3, the *average product* (AP) is, in this case, the technological efficiency of the machine. Column 4, the *marginal product* (MP), is the change in output per unit change in input. Note that this should be plotted between the lines of input data as

$$\frac{\text{col. } 2(n+1) - \text{col. } 2(n)}{\text{col. } 1(n+1) - \text{col. } 1(n)}$$

Had the machine had a wide range of operating characteristics it might have been that this increased with increasing output before it started declining — a common economic case.

It can be seen that, with this particular motor, after an input of some 125 W, increasing equal units of input result in increasingly smaller increments of MP, although the average product (efficiency) increases up to an input of about 250 W.

The decline of MP, gives an advanced warning of the approaching decline in technological efficiency, AP.

The significance of AP and MP is very important, and it underlies most microeconomic studies.

We can now complete Figure 2.1(b), plotting the relationship between energy input and AP and MP which do not have units since they are derived from products of power divided by power.

16

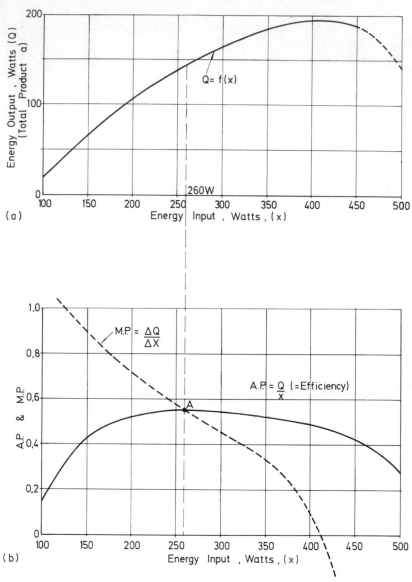

Figure 2.1(a) Total product curve of an electric motor. (b) AP/MP curves of an electric motor. Full curve, AP: broken curve, MP

Point A, at 260 W input with 55% efficiency, being the intersection of the AP and MP curves, is also of the greatest significance. This point is at the maximum *technological* efficiency of the motor, since the AP, Q/x, is at its zenith when $Q/x = \Delta Q/\Delta x$. This follows since Q/x is at its maximum when

$$\frac{\mathrm{d}(Q/x)}{\mathrm{d}x} = 0$$

Table 2.3 Performance of a laboratory motor

Energy Input, x (W)	Energy output, Q (W)	Average product (AP), Q/x	Marginal product (MP), $\Delta Q/\Delta x$
Col. 1	Col. 2	Col. 3	Col. 4
100	15	0.15	
			1.00
150	65	0.43	
			0.80
200	105	0.52	
			0.64
250	137	0.55	
			0.52
300	163	0.54	
			0.40
350	183	0.52	
			0.24
400	195	0.49	
			− 0.16
450	187	0.42	
			− 0.94
(500)	140	0.28	

i.e. when

$$\frac{1}{x}\frac{dQ}{dx} - \frac{Q}{x^2} = 0 \quad \text{or} \quad \frac{Q}{x} = \frac{dQ}{dx}$$

If the input is increased, MP falls below AP, and thus it follows that AP (efficiency) is dropping, and vice versa. It should be noted too, that we are back to the law of diminishing returns. As we add more power input we move to the right of point A; we are achieving *proportionately* less in output than we supply in input, so AP falls.

Finally, from an engineering view point, it is noted that power has to be measured in watts, whether or not it is electrical. In the case of our laboratory motor only the input power was electrical; the output was mechanical and could be put to a variety of industrial uses.

2.4   ANALYSIS OF OPERATIONAL COSTS

The production function $Q = f(x)$ can be extended to $Q = f(xM)$, where M reflects the cost of the machine.

Let us assume that electricity costs two cents per kWh and that the cost of the motor is independent of the power or speed, and is one cent per hour. The motor

Table 2.4 Hourly operational costs of a laboratory motor

Input (W)	Variable cost, VC (cents)	Fixed cost (FC) (cents)	Total cost (TC) (cents)	Output (kW)	Average cost, AC (cents/kW)	Marginal cost, MC (cents/kW)	Average cost, AC (cents)
Col. 1	Col. 2	Col. 3	Col. 4	Col. 5	Col. 6	Col. 7	Col. 8
100	0.2	1.0	1.2	0.015	80		147
						2.0	
150	0.3	1.0	1.3	0.065	20		35
						2.5	
200	0.4	1.0	1.4	0.105	13.3		22.9
						3.1	
250	0.5	1.0	1.5	0.137	11		18.3
						3.9	
300	0.6	1.0	1.6	0.163	9.8		16.0
						5.0	
350	0.7	1.0	1.7	0.183	9.3		14.8
						8.3	
400	0.8	1.0	1.8	0.195	9.2		14.4
						—	
450	0.9	1.0	1.9	0.187	10.2		15.5
						—	
(500)	1.0	1.0	2.0	0.140	14.3		21.4

cost is, by definition, a fixed cost, whilst the power consumed is a variable cost. We can thus compile columns 1 to 7 of the table of *hourly* costs (Table 2.4), the output being obtained from Table 2.3.

Column 6, the average cost (AC), is simply total cost divided by total output for any given level of input (col. 6 = col. 4/col. 5) whilst the marginal cost (MC) is obtained by dividing the *increase* in cost between one input and the next by the corresponding increase in output;

$$\text{col. } 7 = \frac{\text{col.} 4(n+1) - \text{col. } 4(n)}{\text{col. } 5(n+1) - \text{col. } 5(n)}$$

At about 400 W input, MC becomes negative and is meaningless.

For Δ movements of input (in this case 50 W) the marginal data should be plotted between the lines of increment in the table; only when increments become increasingly small and ΔMC becomes δMC can the data be plotted on the same lines as input.

The data from columns 1, 6, and 7 of Table 2.4 can now be plotted as Figure 2.2. At point B, $MC = AC$, and the motor should be operated at this point for maximum *economic* efficiency. Note that this is at an input of approximately 380 W whereas according to Figure 2.1(b), the maximum *technological* efficiency (the peak of the AP curve) was achieved at an input of only 260 W, when, according to Figure 2.1(a), the output was some 140 W.

Let us again establish the significance of marginal cost. If we increase power

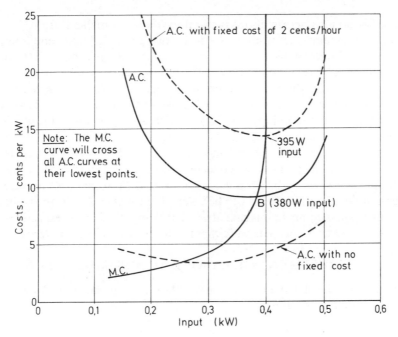

Figure 2.2 Average cost (AC) and marginal (hourly) cost (MC) curves for a laboratory motor. Note that the MC curve will cross all AC curves at their lowest points

inputs to produce to the right of point B, the extra cost of producing more units will be greater than AC, so it follows that AC is lowest at the point of intersection, B.

If we consider point B, where the input is 380 W, reference to Figure 2.1(a) shows that the output is some 190 W. This means that with such machines at a fixed cost of motor usage of one cent per hour and an electricity cost of two cents per kW h, if 140×190 units of power per hour were needed, then *minimum cost* would be achieved by using 140 machines, not 190 which purely technological considerations would dictate.

In considering why there is a difference between technological and economic efficiency we must consider marginal costs again. Referring to column 7 of Table 2.4, for inputs exceeding some 400 W, the marginal cost, which was previously increasing, becomes negative, and meaningless since increasing inputs result in reduced outputs.

The marginal cost is not related to fixed cost, only to variable cost. From Table 2.4 we notice that MC is increasing for, with constantly increasing costs of input (electricity consumed) the output is not increasing *pro rata*. Thus marginal cost is increasing, irrespective of the fixed cost and the *amount* of the variable cost. However, at a given point, in this example approximately 400 W input, extra input is giving almost no extra output, so the marginal cost becomes infinite. It is meaningless not only mathematically but in the real world, for no engineer would increase his inputs (factors of production) to obtain a smaller output!

If we now assume the motor costs to be two cents per hour and complete column 8 in Table 2.4 for a fixed cost of two cents per hour, MC (being independent of FC) remains unaltered whilst the AC curve shifts upwards and to the right, the economic input now approximating to 395 W. Conversely, if we plot the AC curve for zero fixed costs, it moves downwards and to the left crossing the unaltered MC curve at an input of 260 W, which is the position of the maximum technological efficiency as can be seen from Figure 2.1(b). Thus it can be concluded that irrespective of the amount of the variable costs (in this case the cost of electricity) the point at which technological and economic efficiencies are coincident occurs where the FC is zero; however, as FC (cost of plant) increases, higher input (and output) levels will be justified (the higher the FC the higher the input) up to a maximum (in this case about 400 W input) where, irrespective of the size of the fixed cost, the economic input will remain almost unaltered.

2.5 ADJUSTMENT OF LABOUR TO MINIMIZE COSTS

Let us now consider the employment of labour in production. Now output $Q = f(L)$, where L is labour.

Table 2.5 Labour cost of production of ashtrays

Units of labour, L	Labour cost ($)	Total output, $TP = Q$	Average output $AP = Q/L$	Marginal output $MP = \Delta Q/\Delta L$	AC ($)	MC ($)
Col. 1	Col. 2	Col. 3	Col. 4	Col. 5	Col. 6	Col. 7
0	0	0	0		0	
(0.5)				29		0.34
1.0	10	29	29		0.34	
(1.5)				43		0.23
2.0	20	72	36		0.28	
(2.5)				51		0.20
3.0	30	123	41		0.24	
(3.5)				53		0.19
4.0	40	176	44		0.23	
(4.5)				49		0.20
5.0	50	225	45		0.22	
(5.5)				39		0.26
6.0	60	264	44		0.23	
(6.5)				23		0.43
7.0	70	287	41		0.24	
(7.5)				1		10.00
8.0	80	288	36		0.28	
(8.5)				−27		
9.0	90	261	29		0.34	—
(9.5)				−61		
10.0	100	200	20		0.50	

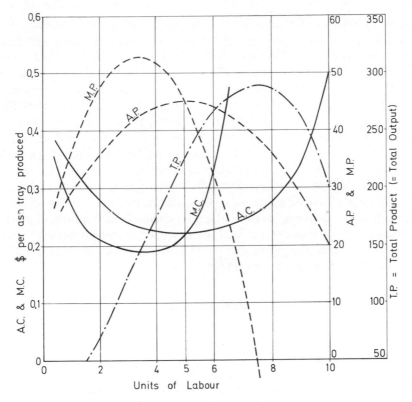

Figure 2.3   Optimization of production of ash trays

Consider the case of a factory with adequate buildings, machinery, etc., where we wish to find the quantity of labour which minimizes the average cost of units produced. This simple case applies to that of a farmer who is producing one given crop from given land and needs to optimize his labour force; but, as engineers, let us consider the production of, say, ashtrays.

Let us assume that $Q = -L^3 + 10L^2 + 20L$ (where Q is the number of ashtrays produced per day and L is the number of units of labour employed, each unit costing $10 per day)*. This production operation will result in Table 2.5, where columns 4 to 7 may be calculated. The average output, in economic terms, is called the *average product* (AP) and is the output divided by the units of production (in this case labour). Similarly, the marginal output is called the *marginal product* (MP) and is the rate of change of output divided by the rate of change of input. In the same way the total output is called the *total product* (TP). The data from Table 2.5 can be plotted as in Figure 2.3. MP will equal AP at

*It may be queried why $Q = -L^3 + 10L^2 + 20L$. In the real world such a function could not be postulated, but the engineer would adopt a deductive process and column 3, the output, could be determined for given sizes of the labour force, and a regression function derived to interrelate labour and output. Many engineers would use graphical rather than regression methods.

the same number of units of input as MC equals AC since, by definition, an increase of input will result in relatively less output, per unit of output, and the AP curve must thus fall. The TP curve itself will, of course, peak at greater units of input before it falls, according to the law of diminishing returns. Note that the AP/MP curve intersection, and that of the MC/AC curves, is at five units of labour. If more, or less, than five men are hired, AP will fall, and AC will rise. It is also worth noting that at about 7.5 units of labour, MC becomes almost infinite. As in the last example, since fixed costs (in this case enterprise, land, capital) are apportioned to production costs, the AC curve will rise, but the MC curve will remain unaltered. Thus the optimum labour (when $MC = AC$) will be more than five units, the maximum total product (total output) being achieved at 7.5 units. This latter point is of significance only when costs are of secondary importance, as in war time in the short run; in normal times we would hire five men, to produce 225 ashtrays a day at a *labour cost* of 22 cents each, or six men producing 264 at 23 cents, or seven men producing 287 at 24 cents, to which costs the *reducing proportion* of fixed costs must be added.

PROBLEMS

Problem P2.1

A contractor has to pour 37.5 m^3 of concrete per day. His labour rate is $40, and mixer hire $20 per day. Materials cost $10 per m^3. As is to be anticipated, he can obtain economies of scale, and the law of diminishing returns applies to the productivity, resulting in the following daily outputs in cubic metres per mixer, the number of labourers being in brackets.

$$2(1); \quad 5.5(2); \quad 9.5(3); \quad 14(4); \quad 16.5(5); \quad 19(6); \quad 21(7); \quad 23(8).$$

The AC and MC curves have to be plotted, and the number of mixers and labourers he should hire determined, together with cost, so that minimum cost per cubic metre of concrete is involved.

Problem P2.2

A road surfacing contractor has to hire men at $5 an hour to be in attendance with his plant. By trial and error he has determined the following hourly outputs of surfacing with a unit of plant.

No. of men	2	3	4	5	6	7	8
Hourly output (m^2)	60	125	180	200	212	220	225

Determine the number of men he should hire to achieve minimum costs of production, and the minimum costs per m^2, if he has to pay $50 or $100 an hour for his plant.

Chapter 3

Production in the Long Run

3.1   OTHER FACTORS OF PRODUCTION

The motor we discussed in Chapter 2 can be regarded as capital, and we have illustrated the case of optimizing labour. We have also shown, for example, that ashtrays were cheapest in labour costs if we produced 225 a day; but what if the demand were for, say, 405? In this case we would adjust not only the labour but also the other factors of production: enterprise, land, and capital. However, whilst in the short run we can change one factor, e.g. hire or fire casual labour, we cannot change the other factors so quickly — it takes time to raise more capital and extend our factory. This brings us to the study of production in the *long run*.

In the long run, by definition, the producer has time to effect major changes to his production process to adjust outputs and costs by changing all the four factors.

3.2   THE TWO-DIMENSIONAL MODEL

We will discuss here a simple model of changing only two factors simultaneously — say, machinery (K, capital) and labour (L).

Consider a road surfacing contractor, using plant and labour. Then $Q = f(K, L)$, i.e. output is a function of K (capital or plant) and of L (labour). Assume that the greatest technological efficiency results when $K = L/10$, i.e. when 10 men are in attendance on each piece of plant. Assume also that the output of 10 men and 1 unit of plant is 2 units which means that $Q = K + (L/10)$. Thus 20 men and 2 units of plant will give an output of 4, but 20 men and 1 unit of plant will not give an output of 3 since this combination will not satisfy the formula for maximum efficiency, i.e. $K = L/10$.

We can now construct line OA on Figure 3.1. Line OA is a *production line*, along which the contractor can move, increasing K and L in equal ratios. Thus along line OA we can mark various *equal intercepts of output*. For example at point X, $K = 2$, $L = 20$, and therefore $Q = K + (L/10) = 4$ units of output. Suppose the contractor decides to hire more labour, say twice as much, without increasing his plant *pro rata*. Then line OB will become his production line, but at point Y, with 2 units of machinery as before, but with 40 units of labour, he cannot

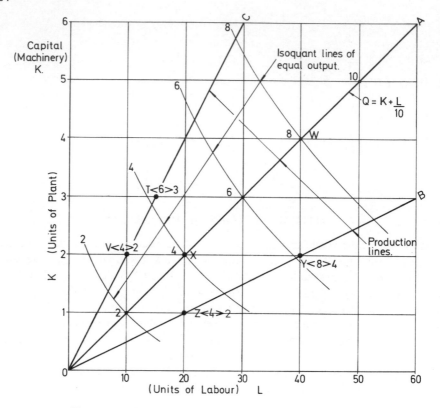

Figure 3.1   Isoquants of output of a surfacing contractor

produce 8 units of output, since, according to the formula for maximum technological efficiency, $K = L/10$, and he would need 4 units of plant to do so. However, as he has 20 more units of labour than at point X he will produce more than 4 units of output. Similarly, as he has less than 4 units of plant, needed at W, he will produce less than 8 units of output. Therefore, at point Y, $Q > 4$ and $Q < 8$. Similarly, his output at Z (with 1 unit of plant, 20 of labour) will be greater than 2 and less than 4.

If, instead of doubling his labour, he halves it, he will produce along line OC and, at point V, with 2 units of capital and 10 of labour he will produce less than 4 units (which requires 20 units of labour) and more than 2 units (which requires only 1 unit of plant for maximum efficiency). Thus point V will have an output of $Q < 4 > 2$. At T it can be reasoned that $Q < 6 > 3$.

3.3   ISOQUANTS

These are lines at any position on which equal quantities, in this case quantities of output, will result.

Having established that at point X output is 4 units, at V output is less than 4 and more than 2 etc., we are in a position to *sketch* isoquant lines, though we have insufficient data to establish them exactly.

It should be remembered that we have assumed, for technological efficiency, that $K = L/10$; this will have been observed in production either by trial and error or from theoretical planning after time-and-motion studies. In undertaking this study we could well have determined exact production output data at such points as Z and T, and could thus have established accurately the location of the isoquants of equal output by conventional contouring techniques.

3.4   PRICE LINES

We have seen that in Figure 3.1 the contractor should expand his production along line OA to achieve *technological* efficiency, according to the function $Q = K + (L/10)$. However, just as in the case of the electric motor, the point of *technological efficiency is not necessarily that of economic* efficiency; so the surfacing contractor may have to depart from this ideal production line to

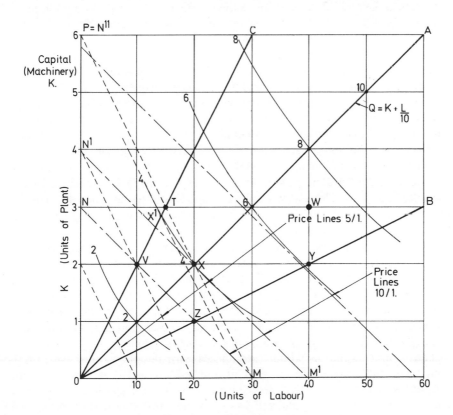

Figure 3.2   Price (isocost) lines

achieve *economic* efficiency, dependent upon the relative costs (per unit of time) of his plant and labour.

This involves the concept of the *price* or *isocost line*. The price line is the locus of points at which equal expenditure is committed on whatever ratio of any two factors of production (in this case capital (plant) and labour) is used. For example, *if* the capital/labour ratio of hourly plant costs to hourly labour costs is 10:1, a series of *parallel* price lines, such as MN, M'N' in Figure 3.2, can be drawn such that OM/ON = 10/1. The isoquants in Figure 3.2 are the same as those in Figure 3.1.

If the labour rate now doubles, we can draw a new series of price lines parallel to MP, such that OM/OP = 10/2 = 5/1. At any point on the new price line MP the same expenditure is incurred; also the more any parallel price line is to the right of MP the greater will be the expenditure.

3.5   SCALE LINES

On Figures 3.1 and 3.2 the line OA is the production line based on the function $Q = K + (L/10)$, and it is the locus we can follow to obtain any given output with maximum technological efficiency.

The locus of a line which joins points through the tangential intersections of isoquants with price lines is called the *scale line*. Isoquants are fixed in position, as discussed in Section 3.3, whilst price lines will vary according to the relative costs of labour and capital (plant). Thus as the price ratio between labour and capital varies, so will the scale lines since the position of tangential intersections of isoquants and price lines will also change.

When the ratio, in any given time period, between plant and labour is 10:1, then price lines will be parallel to MN, and, if he needs an output of 4 units, the contractor will work at point X because, on any given expenditure represented by the price line M'N', if he uses any other plant/labour ratio, his output will decline and his costs will increase. In this case OA is his scale line and production path, for both economic and technological efficiency.

If the labour cost now doubles, the new price lines will become parallel to MP, and the contractor will shift from X to X' to maintain his output of 4 units. His costs will have increased, since X' is to the right of the original price line M'N', but he will be using less (of the more expensive) labour and more of the plant, the price of which is unchanged. OX'C, the locus of points such as X' for differing outputs, is the scale line, i.e. the new production path. On these scale lines $Q \neq K + (L/10)$, though $Q = f(K, L)$, since $K \neq L/10$.

Thus it follows that as labour rates increase (or the plant rates decrease) the contractor will tend to use more plant and less labour to achieve a required output (as determined by the selected isoquants); he will sacrifice technological efficiency to minimize cost, and will do so by working along the appropriate scale line. OB would be his scale line if plant costs doubled or labour costs halved.

3.6   THE SHAPE OF THE PRODUCTION AND
SCALE LINES

In Figures 3.1 and 3.2 we have assumed that the scale and production lines are linear; this infers that there are no economies of scale and that the law of diminishing returns does not apply.

However, these lines will only be linear if the isoquants are parallel to each other on their chords, which will not necessarily be the case. In practice the entrepreneur will establish his isoquant lines from experience or by time-and-motion studies.

3.7   ADJUSTMENT OF PRODUCTION INPUTS TO
OFFSET CHANGING COSTS

As demand increases the entrepreneur will produce more, usually at higher costs, but in the long run he will change his other inputs to minimize these. Similarly as plant costs or labour costs change, he will employ more, or less, to maintain production at lowest costs.

This is best explained by an example.

Example 3.1

Assume that an entrepreneur who manufactures motor cars has established the isoquants in Figure 3.3. On each isoquant, with varying ratios of machinery to labour he can obtain equal outputs. Assume that his machinery hire charges are $100 per machine, and his labour costs are $20 per day.

Now the ratio of machinery costs to labour costs is $100 to $20, or 5 to 1. We must now draw a series of parallel price lines, labelled A, at this slope. For example, if a line passes through 20 units of machinery and 0 units of labour, the cost will be $2000 a day; such a line will thus pass, at zero units of plant, through $2000/$20 = 100 units of labour.

Now suppose the entrepreneur wishes to produce one and a half cars a day, then he can work at any point on the 1.5 isoquant. If follows that, to reduce costs, he will work on this isoquant at the point where it touches the lowest price or isocost line, at which point it will be tangential. This is at point M.

We can now construct the scale line OAA through the tangential contour points such as M on all isoquants — which will be the scale line or production path. It can be seen that to produce 1.5 cars a day 16 units of machinery at $100 a day, and 70 units of labour at $20 a day will be employed, at a total cost of $100 × 16 + $20 × 70 = $3000 a day, or $2000 a car.

Note that, in this case, the scale line OAA is not linear; neither are the isoquant intercepts equal — the shape will be affected by economics of scale and the law of diminishing returns.

Suppose, now, that the labour rate drops to $10 a day. In this case we should

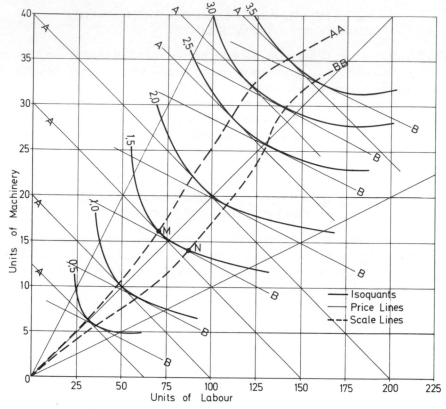

Figure 3.3 Manufacture on motor cars in the long run

construct a new series of price lines, labelled B, at a price ratio of 10 to 1. By similar construction we can obtain a new scale line, OBB, and to make 1.5 cars a day the entrepreneur will shift to N, increasing his labour force from 70 to 88 men, and *reducing* his machinery from 16 to 14 units.

At N his new production cost will be $100 × 14 + $10 × 88, $2280 or $1520 per car. Note that if he had remained at M his costs would have been $1000 × 16 + $10 × 70 = $2300 or $1535 per car (against costs of $2000 per car with the higher labour rate).

From this example it can be seen clearly that as the cost of one factor of production drops, *more of that factor will be employed*, and *less of another*, for any given output. This explains why if, as a result of a union dispute, labour rates rise, less labour and more machines will be employed, and costs will not rise proportionately with labour costs for constant output.

3.8 AVERAGE COST IN THE LONG RUN

In Figure 3.4 a series of short-run average cost curves, of the type derived in Figure 2.2 and 2.3, labelled A, B, C, and D, are illustrated. In the short run a firm

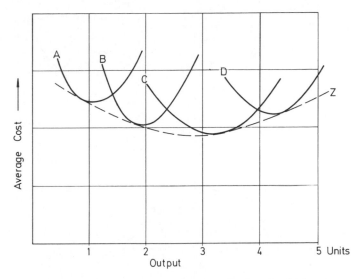

Figure 3.4   Average cost/output curves of a firm in the long run

can adjust other inputs, i.e. factors of production, to operate on other curves such as B, C, and D. Curve Z is the envelope of such curves and is therefore the long run average cost (LRAC) curve. It can be seen that the LRAC curve will have the same concave shape as the SRAC curves, but it will be flatter and altered in position. An average cost curve is the curve of production costs against output, and is thus a production curve, *but it is not the supply curve.*

3.9   MARGINAL COST IN THE LONG RUN

Figure 3.5(a) illustrates the long-run marginal cost (LRMC) curve, which crosses the LRAC curve at its nadir.

The long run marginal cost is defined as the addition to total cost attributable to an additional unit of output, when all units of input (as is possible in the long run) are adjusted optimally. At point E, long-run marginal cost will equal long-run average cost; at one extra unit of output the *AC* will rise and it follows that the *MC* will rise even more and vice versa, and hence the LRMC curve will cross the LRAC curve as illustrated.

3.10   TOTAL COST IN THE LONG RUN

The long-run total cost (LRTC) curve (Figure 3.5(b)) can be derived from Figure 3.5(a).

Total cost is the product of average cost and total output. Thus, the total cost at *any* point A of Figure 3.5(a) is the product of average cost Oc, and output Om.

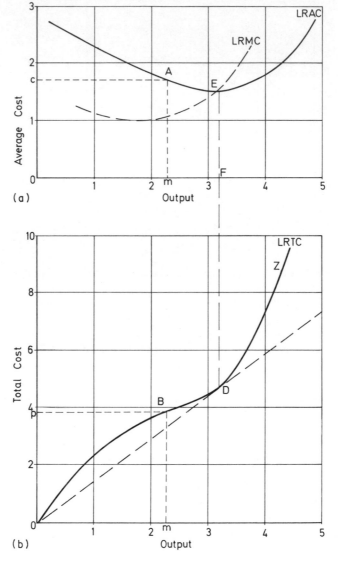

Figure 3.5(a) Long run average and marginal costs of a firm;
(b) long run total costs of a firm

Thus in Figure 3.5(b), at output O*m*, the total cost will be O*p* = O*m* × O*c*. This will result in the S-shaped curve OBDZ.

If a straight line OD is drawn through the origin O, it will touch the LRTC curve (OBDZ) at D, and D will lie under E, in Figure 3.5(a), where the LRAC crosses the LRMC curve. Of course the straight line OD would be the LRTC curve *if* costs were independent of output, and *each* item cost O*c* in Figure 3.5(a).

PROBLEMS

Problems P3.1

A pump manufacturer has determined his daily output of pumps to be as follows:

Daily output of pumps	No. of machines No. of men						
1	$\dfrac{5}{10}$	or	$\dfrac{3}{21}$	or	$\dfrac{2.5}{30}$	or	$\dfrac{2}{50}$
2	$\dfrac{7}{30}$	or	$\dfrac{5}{34}$	or	$\dfrac{4}{40}$	or	$\dfrac{3}{70}$
3	$\dfrac{10}{50}$	or	$\dfrac{8}{55}$	or	$\dfrac{6}{70}$	or	$\dfrac{5}{100}$
4	$\dfrac{12}{90}$	or	$\dfrac{11}{92}$	or	$\dfrac{10}{100}$	or	$\dfrac{9}{120}$

His production paths must be determined when labour costs $20 per day, machines $200 per day, and when machinery costs drop to $100 per day.

Problem P3.2

Determine the unit cost of production for an output of 4 pumps a day for the two conditions in Problem P3.1.

Problem P3.3

Daily overheads are $300, materials cost $100 per pump, and demand is for 4 pumps at $1000, or 3 at $1100, or 2 at $1200 each. Determine the maximum daily profit achievable in Problem P3.1 when machinery costs $100 and labour $20 a day.

Chapter 4

Derivation of Supply and Demand

4.1 SUPPLY — THE PRODUCTION GOAL

We have seen in Figure 3.5, that in the long run, as in the short run, with increasing inputs and outputs, average costs will reduce until a minimum cost is achieved (at which point LRAC = LRMC) after which increasing outputs will result in higher average costs. If the producer can sell his products at a given price above the average cost he will make a profit, and (provided that the demand is still there) if he is in the downward sloping position on the AC curve and can reduce average costs with increasing output, he will increase his production to increase his profits. Indeed, even on the rising part of the AC cuve, he will increase his production beyond the point of minimum average cost (to some point we will determine later), and will work on the part of his AC curve where average costs are rising. In fact, the higher the sales price he can command, the more he will increase his output, and hence the total income, though not necessarily his *percentage* profit.

Thus the supply curve of price (vertically) to quantity (horizontally) will rise from left to right as in Figure 4.1(a). In the real world, if we are dealing with small changes in output, i.e. S_1 to S_2 we can simplify the supply diagram to a straight line, as in Figure 4.1(b).

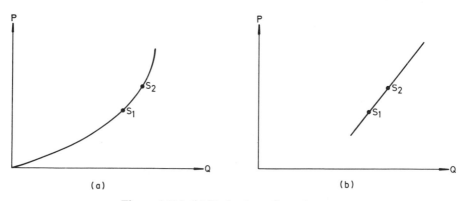

Figure 4.1(a), (b) Derivation of supply curves

4.2   DEMAND — THE LIMIT OF PRODUCTION

Before we can examine the price question in detail we must consider demand, the relationship in the market between price and the quantity in which buyers are interested. The producer exists to maximize his profits; he can work to reducing or increasing production costs, but he cannot calculate his output, revenue, or profit, until he knows the market price, which is a function of demand.

4.3   THE CONSUMER'S MOTIVATION

In economic terms the consumer seeks to maximize *utility*, just as the producer seeks to maximize *profit*.

Utility is an expression, or measure, of human satisfaction or pleasure or welfare. The consumer buys butter for its calorific value or taste, whilst he buys a car for the convenience in driving it; these are referred to as *utility*. If he buys one car it may give him a utility of x, two cars will give him less than $2x$, until his 'nth' car will give him no additional satisfaction, or utility, so he will not buy it. Indeed, he may well decide to spend his money instead on enlarging his house. In economic terms he will gain more utility from improving his house than buying another car.

There are four basic factors of production; likewise there are many factors which motivate the consumer. The buyer, albeit subconsciously, seeks to maximize the utility of his spending power, and is faced with an infinite number of decisions as to how to allocate his resources. In economic analysis it is convenient to consider two variables only, just as we created a two-variable model in the case of long-run production. We can consider, for example, the exchange of money (in itself valueless, but capable of purchasing future utility if it is conserved) with say, a motor car; or, given the commitment of spending, not saving, money, we can consider its allocation between, say, whisky and beer. This leads to the concept of indifference curves.

4.4   INDIFFERENCE CURVES

These represent the locus of the consumer's indifference as to how many units of W (say, whisky) and of B (say, beer) that he purchases (and consumes). With a given income he may be indifferent as to whether he has 10 units of B and 5 of W or, say, 8 units of B and 6 of W.

We have to determine the shape of his indifference curves. Firstly we assume (incorrectly) convex curves. Let Figure 4.2(a) represent his indifference, and I_1 his convex indifference curve with a given income available for expenditure. I_2 will give more utility than I_1, I_0 will give less.

If his income (or expenditure) is reduced he can have less of either or both, and he moves to indifference curve I_0; conversely if it increases he moves to I_2, and will buy more of either or both.

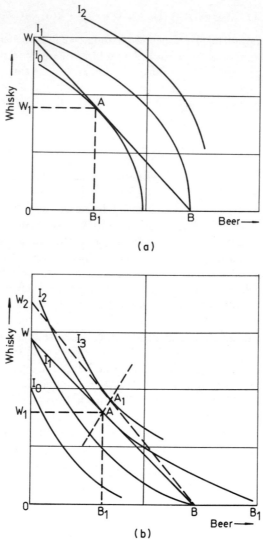

Figure 4.2   Indifference curves: (a) postulated;
(b) actual

Assume that the consumer's available income, I, permits him to buy $0W$ units of whisky at price P_W, *or* $0B$ units of beer at price P_B. Then

$$I = 0W \times P_W = 0B \times P_B$$

Therefore

$$\frac{0W}{0B} = \frac{P_B}{P_W}$$

We can thus draw a line WB, called the *budget line*, anywhere along which the consumer can spend his total available income. As with the price line in production he can spend the same money anywhere on this line. In other words he can, at a given income, buy $0W$ units of whisky, and no beer, $0B$ units of beer and no whisky, or $0W_1$ whisky and $0B_1$ of beer etc. To maximize his utility, the consumer wishes to adjust his expenditure to lie on the highest indifference line; at the same time he wishes to balance exactly the loss of utility in sacrificing a unit of one good for a unit of another (this is known as the *marginal rate of substitution*). He will achieve this condition when the budget line is tangential to the highest indifference curve. In Figure 4.2(a) it is tangential at point A on the lowest indifference curve (I_0), so it follows that indifference curves are *not* convex upwards.

In Figure 4.2(b) with concave indifference curves, the budget line will be tangential to a higher indifference curve at point A on the I_2 curve. It thus follows *that indifference curves are always concave upwards** as in Figure 4.2(b) when the consumer will buy $0W_1$ units of whisky and $0B_1$ units of beer for his utility will then be at its maximum, I_2, by definition, having more utility than I_1 or I_0.

Let us now consider what will happen if the price of beer remains constant, whilst that of whisky drops, available income remaining the same.

Let the new maximum consumption of whisky be $0W_2$ and of beer $0B$, as beer cannot alter.

The budget line is determined by income, and thus will move in Figure 4.2(b) from BW to BW_2 where

$$0W_2 = \frac{0B \times P_B}{P_{W_2}}$$

The budget line (still representing the same available income) will now touch a higher indifference curve I_3 at A, and it can be seen that the consumer can now

---

* The concavity of indifference curves can be explained as follows.
The consumer works to maximize his utility, by using the highest indifference curve. The utility function is given by

$$U = U(B, W)$$

Now an indifference curve is, by definition, a curve along which utility remains constant. Therefore

$$U(B, W) = c$$

where c is a constant. By differentiation we obtain

$$UB\,\delta B + UW\delta W = 0$$

and so

$$\frac{\delta W}{\delta B} = -\frac{UB}{UW}$$

and, as B increases, $\delta W/\delta B$ decreases, that is the slope of the curve decreases or flattens, proving that the indifference curve is concave upwards.

buy more whisky (which is cheaper) *and* more beer (which is not) by moving from point A to point A_1. The line AA_1 will rise from left to right.

By similar reasoning we can see that it will follow a similar locus if the price of beer drops. Of course, with the data available we cannot tell the amount of increase in either product; all we can say is that as price of one commodity drops, more will be bought of both.

This is a two-dimensional model of a multi-dimensional world, so we can say that as the price of one commodity drops, more will be bought of *all* goods, and vice versa. We now consider what happens if the prices of whisky and beer remain unaltered, but available income rises. In this case we will have a new budget line, parallel to but higher than the old one; again we will contact a higher indifference curve and buy more of both goods — of all goods in the real world. Thus it is apparent that as price drops we will buy more, as it rises we will buy less of a commodity (and, indirectly, of other goods), even though we cannot quantify the degree.

The reader should work out for himself the effect of other price moves on either or both products.

4.5 SUPPLY AND DEMAND

We have thus proved that as the price drops the producer will produce less of a commodity whilst the consumer will buy more, and we have also seen that many factors will affect the shape of the supply and demand curves. Let Figure 4.3(a) represent supply and demand curves for an article. Irrespective of the slopes of the S and D curves they will intersect at A, and it is apparent that $0Q_a$ goods will be produced, and sold, at a price of $0P_a$. For small changes of supply and demand we can consider linear functions for both S and D and derive Figure 4.3(b) as our model.

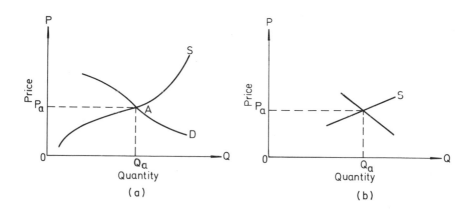

Figure 4.3   Price–Quantity (P/Q) curves

PROBLEMS

Problem P4.1

The daily supply curve of cups is given by the formula $P = Q/1000$, whilst the demand curve is $P = 1000/Q$. Determine the daily demand and price.

Problem P4.2

The cups in Problem P4.1 are so well advertised that the demand doubles. Determine the new price and daily output.

Chapter 5

Determination of Market Price in Perfect Competition

5.1   PERFECT COMPETITION

In Chapter 4 we have assumed that neither the supplier nor the buyer is large enough to affect market prices by his own action; in other words that market prices will not be affected by individual action. This is the principle of *perfect competition* which exists when four conditions are met:

(1) Producers (sellers) and buyers must be of similar size and there must be large numbers of each, so that the action of an individual firm or buyer can have no perceptible effect on the market.
(2) The product must be homogeneous, that is, of the same kind.
(3) All resources must be mobile.
(4) All producers and buyers must have a fair knowledge of the market.

It is obvious that really perfect competition cannot exist in the real world. Monopoly (single producer), monopsony (single buyer — which certainly exists in developing countries) and governmental interferences in the market, which is now universally practised, upset the conditions of perfect competion, but nevertheless, these principles underly much of economic analysis and constitute a firm foundation on which we can build other models to suit other cases.

5.2   DERIVATION OF MARKET SUPPLY CURVES

We have seen in Chapter 3 how in the long run the producer can adjust his factors of production to produce the lowest average cost curve for his firm. He will continue production to maximize his profit until marginal cost equals sales price, so he will operate along the rising part of the MC curve. This MC curve can be regarded as the supply curve of his firm, as long as he has included his overheads.

In Figure 5.1 we can plot the long-run marginal cost (supply) curves for firms A, B, C, etc. The market supply curve will then be the envelope of these individual curves, for the existence of several firms means that more goods can be produced for a given price.

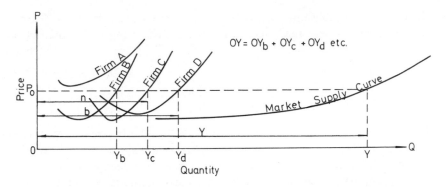

Figure 5.1   Derivation of market supply curve

The market supply curve can be constructed since, at any price P_0, $0Y$ units of goods can be produced, where $0Y = 0Y_b + 0Y_c + 0Y_d + \cdots$, where firms B, C, and D respectively can produce $0Y_b$, $0Y_c$, $0Y_d$, etc. at the given price. Note that in Figure 5.1 the less efficient firm A cannot add to the market supply curve at price P_0, but will do so when the market price rises.

Just as we have constructed a market supply curve by adding the LRMC curves of individual firms, so we can construct a similar market demand curve by summing the demand curves of individual buyers.

5.3   ELASTICITY

We repeat that both the market supply and market demand curves will not be linear, but over *small* ranges we may conveniently consider them to be so, and can thus deduce Figure 5.2 where S is the *market* supply curve and D the *market* demand curve. The slopes of these curves are measures of their elasticities; *elasticity* is a parameter measured by economists as an aid to economic forecasting.

Engineering elasticity is the change in stress with an incremental change in

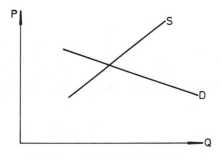

Figure 5.2   Price–quantity curves

strain. It is a recoverable amount, and as strain is relaxed, so is the stress. A similar concept applies with elasticity in its economic sense, being the change of one parameter with the incremental change in another. The price elasticity of demand can be expressed as

$$\frac{\delta Q}{\delta P} \cdot \frac{P}{Q} = \frac{\delta Q/Q}{\delta P/P}$$

If the S curve is steep it is called inelastic, because the quantity being supplied will not react much to price changes. A perfectly horizontal supply curve is infinitely elastic, since an increase in demand will not result in a change in price.

Similarly, we have demand elasticity, a vertical demand curve being infinitely inelastic; i.e. a fixed quantity is required irrespective of price.

5.4 EFFECT OF DEMAND AND SUPPLY IN PERFECT COMPETITION

Figure 5.3 illustrates a series of demand and supply curves. Figure 5.3(a) shows the classical case of price determination. The intersection of the market supply curve (S_1) and the market demand curve (D_1) determines both the quantity required and the price. Note that the product $p_1 q_1$, or the area of the rectangle $Op_1 Bq_1$, is the total revenue to the producers, being the product of *average revenue* (AR, average price) and quantity.

In Figure 5.3(b) the demand curve has shifted upwards from D_1 to D_2. This could result if there were a tax reduction and more money became available for consumption, or if an advertising campaign were successful. Note that as the quantity demanded increases, so does the price (q_1 to q_2 and p_1 to p_2). It is interesting to postulate the effect on the cost of living.

In Figure 5.3(c) the demand has remained unaltered at D_1, but the supply has increased in price from S_1 to S_2 (this could be due to a price increase of raw material). In this case the price *increases* from p_1 to p_2, but the quantity *reduces* from q_1 to q_2. Whether or not the total revenue, $p_1 q_1$, has increased or decreased to $p_2 q_2$ depends on the slopes, or elasticities, of the S and D curves.

Figure 5.3(d) shows the case when both the D and S curves rise. P will increase; whether Q increases or decreases will again depend upon the elasticities of the S and D curves.

The S_3 curve in Figure 5.3(e) illustrates the case of an *inelastic* supply — the producers will supply neither more nor less of any product, whatever price is offered. This case of inelastic supply can occur with perishable goods, such as milk (assuming that no processing facilities were available), since, if a supplier finds himself with q_1 litres of milk one day, it will be valueless the next, and he must sell it irrespective of price. If the demand curve is D_1 he will obtain price p_1, if the demand curve drops to D_2, he must then sell at p_2, the quantity q_1 remaining unaltered. The supply of crude oil can be regarded as inelastic in the long run, as the supply cannot be increased. If the demand for oil power increases, the demand curve will shift to the right, and the price will increase without any change in

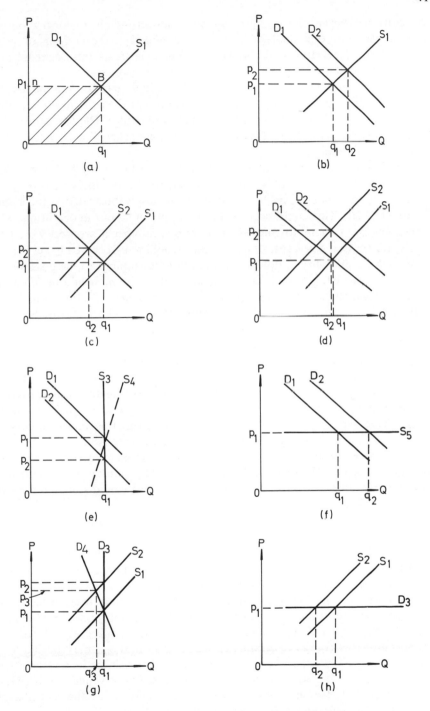

Figure 5.3   *P/Q* curves for the market in perfect competition

quantity. However, the grouping of oil producers tends to form monopoly conditions, and our analysis is strictly confined to conditions of perfect competition. Nevertheless the supply of oil is inelastic as extra demand prices cannot increase the long-run availability of fixed resources.

Note that since revenue is the product of price and quantity, and the supply of the oil is inelastic, suppliers would be foolish to cut their price deliberately, as long-term quantity is constant and revenue would reduce. They can store oil and should only cut prices when they have sold on the market all it can absorb at the prices then prevailing.

The S_3 curve in Figure 5.3(e) is completely inelastic; in the real world the S_4 curve is more likely to apply, and is also regarded as an inelastic supply.

The other extreme case is the infinitely elastic supply illustrated by the S_5 curve in Figure 5.3(f). This could occur with the supply of labour in conditions where there was great unemployment, and an infinite number of workers would offer their services at a going rate of p_1; but they would not work at a lower rate due, say, to the availability of dole or, in a developing country, to the possibiility of living at a subsistence level on their land. In this case q_1 would be hired at p_1 wages, unless the entrepreneur (in this case a consumer of labour which is the supplier) found an increased market for his products and, to increase his output decided to make more goods, in which case his demand curve would shift to D_2, and he would engage q_2 labourers, still at the going rate of p_1.

The D_3 curve in Figure 5.3(g) illustrates the case of infinitely inelastic demand. An electric railway may require q_1 units of power to run its services, and at a supply curve S_1, would pay p_1 per unit. If the supply of electricity increased to S_2 the railway must still have q_1 units, and as the supply curved shifted to the left, would have to pay p_2 per unit of power. (In the long run, this railway might find it economic to switch to diesel power, but this is another issue.)

In practice the extra cost of power would cause fares to rise, which would marginally reduce the demand for electricity, the D_3 curve would shift to D_4 and the consumption would be q_3 at a price p_3; this would still be regarded as inelastic, being a steep curve.

Note that in conditions of inelastic demand, with rising prices, pq, being the supplier's revenue, always increases.

Figure 5.3(h) illustrates infinitely elastic demand. This will occur when substitutes are available and the consumer is indifferent to the product he consumes. If S_1 is the supply curve of diesel fuel, say, to a factory which buys q_1 units at price p_1, and the management is also able to substitute crude tar as a fuel at equal cost, then if the cost of diesel production and supply rises, and the supply curve shifts to the left (S_2), the factory (the consumer in this case) will reduce its diesel purchases from q_1 to q_2, still at price p_1, and will substitute crude tar for the shortfall. In this case the oil firm, as the supplier, has reduced its revenue to p_1q_2 whereas, in Figure 5.3(g), the supplier had increased his revenue to p_2q_1. Thus it can be seen that revenue will be a function of elasticity of demand, as well as that of supply; this introduces the principle of *cross elasticity*.

5.5   CROSS ELASTICITY

Unitary elasticity of demand occurs when a shift of price is counterbalanced by a shift of quantity so that the total revenue remains unchanged. The same principle applies to supply elasticity. What happens to P, Q, or revenue in a market depends on the cross elasticity of supply and demand. Of course, it is not possible to calculate theoretical values of elasticity for supply or for demand, but these are studied in the real world by economists who measure the effects of price changes, and thus it becomes possible to extrapolate the effect of further changes in demand or supply.

Again it is stressed that elasticity is not constant over the full length of a supply or demand curve, but is a variable; it is only constant within practical limits, over a short range.

PROBLEMS

Problem P5.1

Perfect competition exists in the paint industry; the supply and demand data are as follows:

Litres per day	200	300	400	500	600	700	800
Market demand price ($/litre)	4.0	3.2	2.5	2.0	1.5	1.1	0.9
Market supply price ($/litre)	2.0	1.3	1.1	1.1	1.3	1.8	2.5

Government decides to rezone and move the industry away from the market, involving additional transport costs of 50 cents/litre. The market price and quantity produced and sold has to be determined before and after rezoning.

Problem P5.2

In the conditions of Problem P5.1 we have to consider the economics of one firm if it alone had to be rezoned.

Chapter 6

Determination of Output, Revenue and Profit

6.1 OUTPUT IN PERFECT COMPETITION

In Chapter 5 we have been determining price in perfect competition, from the intersection of PQ curves.

As has been stated, the action of a single supplier under perfect competition, by definition, cannot affect the market. For example, in a market of a hundred firms, if one firm raises its prices, it cannot affect the market price. If it sells its products at too low a price, it will reduce its profits; if it tries to sell at too high a price it may not succeed in selling its output. Of course, at too low a price, it can be argued that the firm could increase its production, and hence its total revenue by under-cutting the market. However, this is then no longer perfect competition, for the single firm is thus affecting the market.

6.2 DETERMINATION OF A FIRM'S DEMAND CURVE

In Figure 6.1(a), the market price, p_1, is determined from the intersection of the market demand and supply curves. Since, by definition, no single firm can affect the market price, it can sell all of its production at price p_1. Thus in Figure 6.1(b), we can deduce the horizontal, i.e. infinitely elastic, demand curve for the *individual* firm at this price p_1.

We are now in a position to consider revenue. Since the demand curve is in effect the firm's sales curve, the *demand price is the average revenue* AR.

Also, if we define *marginal revenue* MR as the extra revenue resulting from the production of one extra unit, it can be seen that demand = average revenue = marginal revenue, or $D = AR = MR$, when the firm is at equilibrium.

6.3 DETERMINATION OF A FIRM'S SUPPLY CURVE

Marginal cost is the extra cost involved in producing one extra unit of output. Firms will obviously not be in business at an output where the marginal cost is falling, as they will increase production (to reduce costs) beyond this output level.

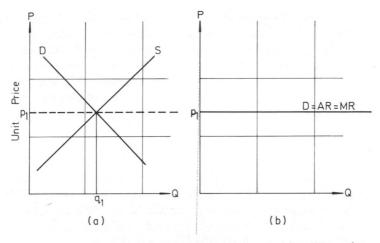

Figure 6.1. Demand Curves under perfect competition: (a) market demand and supply curves; (b) determination of an individual firm's demand curve

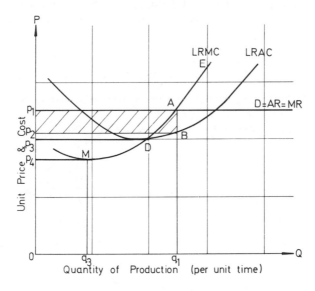

Figure 6.2   Determination of individual firm's long-run supply curves

This was illustrated in Figure 2.3. They will thus produce along the rising part of the marginal cost curve, firstly until it meets the average cost curve (beyond which point the cost of an extra unit of production will raise average costs), then further until such time as no profit is shown by producing an extra unit of production; that is to the output where marginal cost equals the marginal revenue or demand price. At this output it again follows that $MC = D = AR$ when the cost of producing the last unit is the sales price.

Of course, Figure 2.3 was for the short run. However, since the firm wishes to maximize business by lowering costs, in the long run it will adjust all factors of production, and long-run curves (of similar shape but different position) will apply. We can thus deduce Figure 6.2 for long-run conditions, and production will continue to $0q_1$ units where $MR = LRMC$. Income will be p_1q_1, the area of the rectangle $0p_1Aq_1$, whilst costs will be p_2q_1 or the area of the rectangle $0p_2Bq_1$. Profits are thus the area p_2p_1AB, or $q_1(p_1 - p_2)$.

The question is now posed as to what happens if the MR line (the price of the intersection of the *market* equilibrium price) is lower than shown?

If the MR line drops from p_1 to p_3, the profits will become zero — the firm can continue to operate, but only at cost. If, however, the price (the market demand price) is established at lower levels, say at M, where MR intersects LRMC at its lowest point, with a price p_4, then revenue will be less than costs, the firm will show a loss, and it will leave the market, the LRAC being higher than the LRMC. This applies to any demand price which is lower than p_3.

It then follows that the firm's supply curve will be its long-run marginal cost curve, but the firm will operate only on the increasing side of it and *on that part of it which is above the average cost curve*; that is for quantities of production greater than p_3D. In other words DAE will be the individual firm's supply curve.

6.4 DETERMINATION OF A FIRM'S OUTPUT BY TOTAL COST ANALYSIS

This is an alternative approach to the method outlined in Section 6.3. The construction is illustrated in Figure 6.3.

The total revenue plot, TR, will be a straight line through the origin, as the

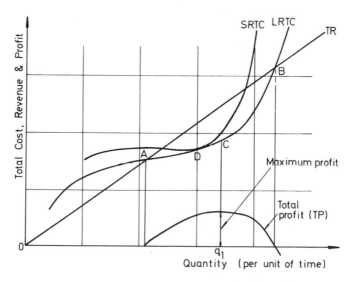

Figure 6.3   Total cost approach to optimization of output

demand price is fixed. The long-run total cost curve (LRTC) will assume the shape shown — at a low output the production costs will exceed the price. At production levels between A and B, total costs will be less than total revenue (or the firm would leave the market) and at levels greater than B costs will exceed demand price, so the firm would not produce. The short-run total cost curve (SRTC) is not important, but will have a similar shape and will lie above the LRTC curve, save at one point D which will represent one of many combinations of the factors of production in the LRTC curve.

On the x axis we may draw a curve for total profit, TP, being the amount that the TR curve exceeds the LRTC curve. The limit of production can be deduced as $0q_1$ where q_1 lies beneath the zenith of the TP curve, and will correspond to point C, the furthest point on the LRTC curve beneath TR, since total profits are maximum at this point.

PROBLEMS

Problem P6.1

There are ten equally efficient firms producing bicycles, each with LRAC curves as follows:

Daily production of bicycles	10	20	30	40	50	60	70
Average cost of each bicycle ($)	53	46	42	40	42	48	57

The daily demand curve is a linear function passing through 400 a day at $60 and 600 a day at $40. Determine the daily production of each firm, together with the daily profits.

Problem P6.2

An eleventh similar firm enters the industry (this is the same as one firm doubling *all* factors of production) in Problem P6.1. The new total volume of sales, together with sales prices and individual firm's profit, is to be determined.

Problem P6.3

Ten similar firms can manufacture motor scooters and have determined that their average prices are as follows:

Daily output	2	3	4	5	6	7	8
Average cost ($)	180	160	145	130	125	130	140

The market demand has constant elasticity and is for 40 a day at $250 and 100 a day at $100. Determine the daily output and sales price of each firm.

Chapter 7

Imperfect Competition

7.1 LIMITS OF PERFECT COMPETITION

It can be argued that perfect competition cannot exist, since one firm leaving the market or altering its prices, can affect the market price. This is true not only in developing countries, where the market is small and suppliers are few, but also in large countries, where specialization results in a few large suppliers. It can also happen with patented goods.

The extreme example of this case is monopoly (but it should be remembered that the reverse case of monopsony — a single buyer — can also occur in small countries where the state, or a cooperative, may be the only buyer). However, true monopoly hardly ever exists as there are usually substitute products, however imperfect, and also there is indirect competition in that all commodities compete for a place in the consumer's budget. For example, if bitumen is required for road construction and there is only one supplier whose price is too high, the engineer can redesign to use tar, or even to use concrete. Time mitigates against the study of monopoly in depth; suffice it to explain the elements of the determination of marginal revenue, its control of profits, and the limitations of output.

7.2 MONOPOLY

In perfect competition the demand curve for the individual firm is horizontal, and $D = AR = MR$, whilst that for the *market* slopes downwards with increasing quantity. In monopoly the demand curve of the one firm is the same as the market's, and it slopes downward, left to right; the demand is the average revenue, but not the marginal revenue, i.e. $D = AR$.

At unitary output $MR = AR$, and since $AR = D$, then $D = MR$, and the monopolist's income would equal his revenue. At higher outputs, which are our concern, since the demand price is falling, his sales price and hence his AR is reducing, and the monopolist will not increase production if his marginal revenue, the increase in revenue for the production of one extra item, is not falling faster than his average revenue. Hence $MR < AR$. Putting it another way, this follows since the demand curve is downward sloping, and the price he receives for his second article is less than that for the first.

7.3 CONSTRUCTION OF THE MR CURVE

In Figure 7.1(a), since the demand price for successive articles reduces, total revenue will increase with quantity at a *reducing* rate, reaching a maximum at Z with q_1 units, after which it will fall to zero with zero price, at q_2 units.

In Figure 7.1(b) the demand curve is shown, and the total revenue TR at any level of production is the product of P and Q. At price OA it is the area OACF. The marginal revenue is decreasing to zero at q_1 (the price beneath Z) and extra units of production will result in reducing revenue. For simplicity's sake we have to assume linear functions for the MR and AR curves, just as we have previously assumed linear functions for S and D curves, which is true for small changes.

Let us consider any output OF on Figure 7.1(b). At this level of output the price paid is AO = FC and the total revenue is the area of rectangle OFCA. Again marginal revenue is the addition to total revenue attributable to the addition of one unit of output (per period of time). After the first unit it is less than price. It follows from this definition of marginal revenue that the total revenue corresponding to any quantity is the sum of all marginal revenue figures up to that quantity. This means that total revenue is the area under the marginal revenue curve (area PFEG). Therefore

$$\text{area OFCA} = \text{area OFEG}$$

Subtracting area OFEBA which is common, area ABG = area BCE. Thus

$$AB = BC$$

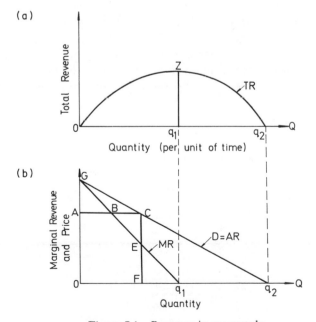

Figure 7.1   Revenue in monopoly

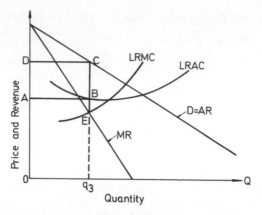

Figure 7.2   Output and profits in monopoly

We now have the tool for constructing the marginal revenue curve. To do so we make $0q_1 = q_1q_2$ and Gq_1 becomes the marginal revenue curve. Of course, if the D (= AR) curve is non-linear the calculation of the MR curve becomes mathematically more complex, but remember, although the model is of the whole, we are concerned only with small changes justifying the linear concept.

7.4   CALCULATION OF OPTIMUM OUTPUT AND PROFITS

As in the case of perfect competition, to maximise his profits, the monopolist will still have to work to the condition that MR = MC,* which occurs at point E in Figure 7.2, the output being $0q_3$. At this level of output his total costs will be similarly the area $OABq_3$, so his total profits, which by definition mean that LRMR − LRMC will be a maximum, will be represented by the area ABCD.

If the LRAC curve lies above the D = AR curve, the area ABCD becomes negative and then the monopolist can only produce at a loss — a situation which is contrary to public opinion which accepts, erroneously, that the monopolist can dictate prices. Indeed, it is possible to conceive of this happening in nationalized

---

* The mathematical proof that MR = MC is as follows.
  Let A be the monopolist's profit, B his total revenue, and C his total costs. Thus

$$A = B - C$$

If Q is the output where profit is a maximum

$$\frac{\delta A}{\delta Q} = \frac{\delta B}{\delta Q} - \frac{\delta C}{\delta Q} = 0$$

but

$$\frac{\delta B}{\delta Q} = MR \quad \text{and} \quad \frac{\delta C}{\delta Q} = MC$$

Therefore MR = MC at the output resulting in maximum profit.

industries, when production could only continue with a state subsidy, either to sale price or to the factory itself. Such a subsidy could even be indirect and could take the form of writing off right-of-way charges from a railway, which would reduce its LRAC. A subsidy of this kind might be justified either in order to keep labour employed, or to maintain the railroad in operation. The railroad might be kept in operation either for strategic purposes or to restrict the use of foreign currency which might be necessary for the extra imported oil needed to run the enlarged road transport industry if the railway were closed.

PROBLEMS

Problem P7.1

Petroleum suppliers have formed a consortium for the supply of diesel oil, and the industry's daily supply characteristics are:

Output (Ml/day)	LRAC (cents/1)
3	9.0
5	8.4
7	8.0
9	8.0
11	8.7

The alternative for the consumers is to switch to crude tar, and this results in a demand curve of constant elasticity passing through two points: 7 Ml/day at 12.5 cents and 13 Ml/day at 8 cents.

Calculate the level of production at which the oil consortium should maintain production to maximize profits, and the size of the maximum profit.

Problem P7.2

The OPEC countries, by agreeing on a price, have created monopolistic conditions, *but* the supply price is controlled by the suppliers. Yet the oil supply is limited, and is thus inelastic.

At the same time, with the development of underdeveloped countries, demand is increasing. Illustrate the situation with PQ curves, with a brief explanation, showing the price changes to be anticipated with time.

Chapter 8

Macro Economics: The Business Cycle and National Income

8.1 MACRO ECONOMICS

Macro economics is the study of economic growth, fluctuations, employment, and price levels. It underlies political philosophy, and is thus of interest to every citizen and voter, though few have studied this complicated subject in depth, Many engineers may have little interest or professional contact with macro economics, but those who become senior civil servants, company directors, or consultants to government, will either be involved in decision-making based on macro-economic philosophy, or, at the very least, in implementing such policy. This applies increasingly to those employed in those small or underdeveloped countries, where problems of foreign currency are of day-to-day concern, and for this reason, an abbreviated introduction to macro economics is worthwhile for every engineer, even though he is not likely to be concerned with detailed calculations. The complications of detail are thus omitted, as are Keynesian and post-Keynesian philosophies.

8.2 THE BUSINESS CYCLE

The business cycle is perhaps unfortunately named, as it infers a cycle of commerce, whereas it is really a cycle of economic well-being, which affects every citizen of the world, except primitives living on a subsistence economy. It is the existence of the business cycle which earned for economics the sobriquet of 'the dismal science'.

The cycle can be plotted with time on the horizontal axis and a 'per cent of long-term trend of economic activity' on the vertical. *Economic activity* can constitute a variety of indicators, from the Gross National Product (GNP),* to consumption of electricity; from percentage employment to expenditure on civil engineering works. The *percentage of long-term trend* is a little more difficult to define. If we are concerned with, say, electricity consumption — which is regarded

---

* GNP is defined as the annual summation of both personal and governmental expenditure on goods and services, plus that of investment expenditure on all new machinery and construction.

as a good indicator of industrial and economic health — only the power consumption *per capita* is really absolute, as the population may be changing. Similarly only the percentage unemployed, rather than the total unemployed, is really of significance. However, most indicators are measured in terms of the national currency, e.g. the dollar, pound, deutschemark, yen or franc, which itself is likely to have a changing value.

Suppose we are concerned with the production of cars, and a country produces a hundred-thousand in a base or reference year at a cost of one thousand dollars each. We can thus say that production of cars was $100 million at *factor* prices. If, in the subsequent year we sold our production for $110 million dollars, we can say that our production was then valued at $110 million at *factor* prices. However, the extra price may have resulted not from more, better, or bigger cars, but from increased costs. Suppose that the costs of indentical production in the motor industry had increased by 6% during the year. Then the value of production in the second year, at base year prices, was only $110/1.06 million, i.e. $103.8 million. Thus we can say that the output of the motor industry had a 3.8% increase compared to the base year. By similar calculations over subsequent years we can establish a long-term trend — which is generally increasing.

Figure 8.1 illustrates the US business cycle for this century, plotted on this basis; we may take the vertical scale as being the percentage of long-term trend of The Gross National Income (GNI).* Note that in 1950, it was less than in 1920 on this basis, though in actual figures it could actually have been higher, since the zero base line was the long-term trend which was patently increasing.

It can readily be seen that GNI was cyclic, if irregular, in nature, and what is true for GNI is true for most other indicators. The highest point of any business cycle, as in 1928, is called a *peak*. This is followed by a *contraction* or *reversal* to a *trough*, as in 1932, followed in turn by an *expansion*. It is interesting to note the peak caused by World War I, the Great Depression of the thirties, with the revival and peak of World War II, and the smaller peak of the Korean War. In the depressions production is at a minimum and unemployment at a maximum, whilst the reverse is true at peaks.

Whilst the business cycle for any one country is national it has international effects, since to a greater or lesser extent, trade is international. For example, if there is a depression in the USA, it will import less chrome, and a depression will be triggered off in chrome exporting countries.

The long-term growth of the National Income explains why it is desirable to plot the vertical scale as a *percentage of the long-term trend* in the construction of a business cycle. Experts disagree as to whether the cycle is a combination of long-, maximum- and short-term waves, but long severe depressions, as occurred in the thirties, are no longer expected as it is considered by many that the application of Keynesian and post-Keynesian economic theories by governments will enable

* GNI is loosely defined as the monetary measure of the annual flow of goods and services. The Net National Product (NNP) is often used as an alternative expression. NNP is the national income plus indirect 'business' taxes which accrue to government. Since GNP can be derived by adding capital consumption to NNP, it follows that all three indices are interrelated and follow broadly similar curves when plotted against time.

54

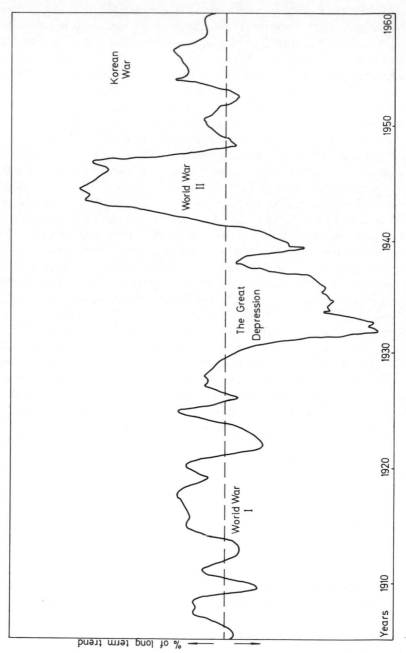

Figure 8.1 A business cycle

them to steer their economies out of a depression. However, a smaller country is less likely to be able to influence its business cycle than a large one, due to international repercussions and the world population explosion, and dropping *per capita* productivity, together with its limited mineral resources, poses new problems.

However, it is an unfortunate fact that war seems to be one certain means of ending a depression, due to the employment of labour and the means of production that it entails, whilst a psychological loss of confidence can trigger off a depression — again, to quote Shakespeare: 'The fault, dear Brutus, lies not in stars, but in ourselves.'

An approach via durable and consumer goods is interesting in that engineers tend to be more concerned with the former, which are more cyclic in their behaviour. Durable goods are loosely defined as those which are long lasting, such as dams, roads, and railway locomotives, and consumer goods those which are quickly consumed, such as food, oil, and electricity.

After a peak, if there is a loss of confidence, then there will be less development planned, and the demand for materials used in the manufacture of durable goods, such as cement and pig iron, will reduce, though man, the consumer, will still continue to eat. Iron and cement workers will become unemployed (but will still eat, albeit on a reduced scale, by living on savings and the dole), and so firstly the durable goods industry will become depressed; then, to a lesser extent, the consumer goods industry. Merchants will reduce their holding of stocks and the contraction is underway.

At the time of the trough, inventories (that is stocks of goods) will be minimal, but durable goods, becoming worn, will need replacement, and the cement and iron workers will be re-employed. With increased income they will start buying more consumer goods, and recovery, i.e. expansion, will be underway. This is a very simple explanation, and it will be shown how fiscal (taxation, etc.) and monetary (availability of money) controls can be used to influence the cycle to iron out the troughs, and also the peaks, which are often accompanied by inflation.

The existence of the business cycle is a fact of life which the engineer must learn to live with, and to which he must adjust. The engineer concerned with production of capital goods in a small firm may have to reduce production, or to diversify production into consumer goods. In a large firm (which, it is assumed, has sufficient capital reserves), production may continue to build up inventory stocks, in the knowledge that these can be sold when 'normality' is reached. However, their sale at the onset of an expansion will tend to offset the recovery, so production should ideally be reduced (using micro-economic principles) to a point where costs do not significantly rise by further reductions in output.

As will be discussed later under cost–benefit analysis, most schemes, particularly in civil engineering, involve large expenditure on durable goods (e.g. on railway works, on roads, and on dams), yet over their long lives they will result in an overall reduction of costs. Such capital expenditure itself may be long term in that it may be committed and spent over several years, over peaks and recessions.

There is a tendency amongst governments to reduce expenditure on such schemes at the onset of reversals, although they are often revived at troughs to generate employment and hence recovery. Logically, however, such long-term works should be proceeded with as soon as a reversal is marked, and, for that matter, withheld at booms before peaks, to iron out the cycle.

Another long-term engineering activity is research and development. Once a decision is made by a firm as to how much it should allocate on research and development, this should continue irrespective of the business cycle. Although such a decision is obvious, it is, unhappily, a sphere of activity which is often axed.

In developing countries, when there is a depression government often reduces its engineering work, with the result that engineering skills will emigrate. Later, when these are required, they are no longer locally available and have to be imported, not only at greater cost, but often paid in foreign currency. The use of local currency to buy foreign currency for such skills means that money is withdrawn from the national economy, delaying or negating a revival.

8.3 THE CYCLIC NATURE OF THE ENGINEERING INDUSTRY

Figure 8.2 illustrates the cyclic nature of the economy of a developing country, although, strictly speaking, only the curves of GNI at 1965 prices and electricity consumption are business cycles, as other curves are not correlated to a fixed monetary base or a 'percentage of long-term trend'. Correction of the 1975 GNI to 1965 prices reduced the GNI from some $2000 million to little over $1400 million, as inflation had occurred; but no comparable correction factor was used for the building and civil engineering prices.

It can be seen that whilst the electricity consumed increased continuously from 1954 to 1975, both the building and civil engineering curves reflect the shape of the GNI curve, but are even more cyclic, and the building curve more so than that of civil engineering.

In this case political confidence occurred in both 1953 and in 1965; a year or two later the expansionary period set in, investemnts planned were comitted, and the economy entered an expansionary phase. With a loss of confidence, in 1957, and with the onset of world recession in 1975, the peaks were reached and recessions began. Private enterprise reduced investment in building, and government had less income to spend on public works at these times.

It is normal practice for many governments to retain sufficient engineering staff to carry the work load at troughs, and to use consultants for the additional work in periods of recovery, with the result that consultants (and architects) will have to rely on continuing through recessions either on savings, or on a few clients in the private sector. Thus their employment will be even more cyclic than the business cycle would indicate. Logically government should employ them on planning during recessions — though this would not help the contractors.

In Figure 8.2 the consumption of electricity showed no major cyclic changes but rather a *status quo* at recessions; so it is reasonable to assume that mechanical

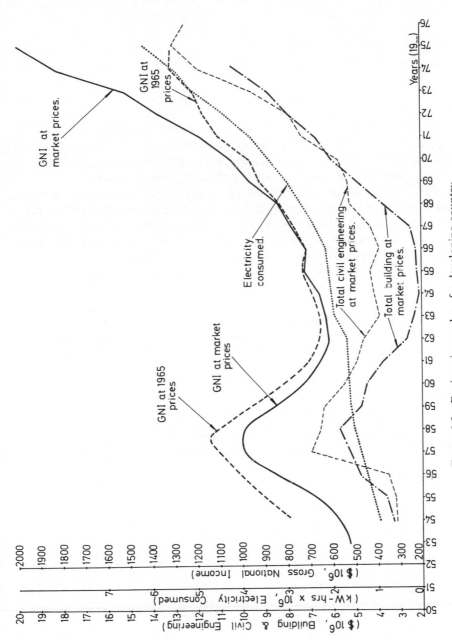

Figure 8.2   Engineering cycles of a developing country

and electrical engineering are less cyclic than civil engineering — probably due to so much of their output being concerned with consumer goods. At the same time large, long-term civil engineering projects, such as hydro-electric schemes, should not be cyclic as construction time is long compared to the cycle, and this, together with essential engineering maintenance, explains why civil engineering is less cyclic than building.

8.4   NATIONAL INCOME

A country's GNI is closely related to its GNP. The technicalities of tis compilation do not concern us, but every country seeks to increase its GNI since this, divided by its population, is a measure of the citizen's standard of living, if it is measured at constant (in this case 1965 is the base year) prices.

Table 8.1 shows how the GNI of a developing country has been increasing every year, save 1966, but in terms of constant (1965) prices it showed a decline in 1966, and since 1974. Thus in each year save these the national income increased, and (assuming that population increased at a lower rate) the average income also increased, together with an increasing standard of living.

The difference between the GNI at market, and at constant prices, is a measure of the inflationary increase in cost of goods and services.

The data from Table 8.1 are plotted in Figure 8.3. Except in 1966, whilst GNI at market prices was increasing each year, the business cycle shows up clearly when it is plotted at constant, in this case 1965, prices. The trough in 1966, and peak in 1974 are most apparent.

Table 8.1   The GNI of a developing country in millions of dollars

Year	GNI at market prices	GNI at 1965 prices	Annual increase in real terms (%)
1962	615	—	—
1963	631	—	—
1964	660	—	—
1965	722	722	—
1966	714	700	− 3
1967	780	712	2
1968	833	809	14
1969	960	924	14
1970	1049	962	4
1971	1180	1072	11
1972	1378	1165	9
1973	1509	1210	4
1974	1829	1322	9
1975	1994	1311	− 1
1976	2110	1259	− 4
1977	2320	1200	− 5

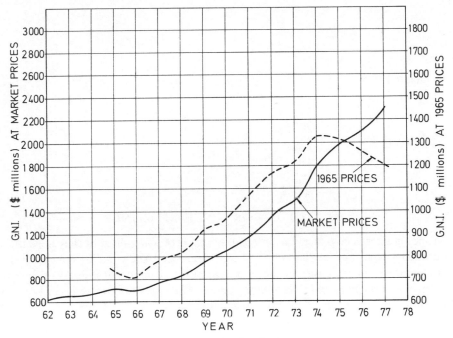

Figure 8.3   GNI of a developing country

PROBLEM

Problem P8.1

The cyclic nature of employment in the architectural, civil engineering, electrical engineering, and mechanical engineering professions should be analysed, placing them in order with brief reasons.

In the civil engineering industry there are contractors, consultants and governmental employers. Give reasons why consulting engineering may be more cyclic than contracting, and both more than the governmental sector.

Chapter 9

Money

9.1   THE IMPORTANCE OF MONEY

Money concerns engineers both privately and professionally, but they need little detailed knowledge of this vast subject. Thus it is sufficient to condense the subject into the principles of the creation and destruction of money, and of its control as a background to monetary policy. Monetary policy itself is important because, together with fiscal policy, it is used as a tool to regulate the evils of inflation and of unemployment, and thus it controls the amount of engineering work.

9.2   THE ROLE OF MONEY

Coins today may cost almost as much in material and labour to manufacture as their nominal values; but since the amount in circulation is relatively small, they are ignored in the study of money. Of course, pure gold and silver coins are no longer used as money.

   Money in the form of notes and cheques, which is the monetary form with which engineers are chiefly concerned, has no intrinsic value, but serves as a medium of exchange to enable the holder to obtain goods from a seller in a ready and convenient manner, without resort to barter, and at a time to suit both buyer and seller. It is the confidence that the seller of an article or service has in the notes or cheque that he receives, to dispose of them at any time he chooses for other goods or services, that gives the value to such money. It is best regarded as a measure of productivity.

9.3   THE CREATION OF MONEY

Money in the form of bank notes is printed by the national government, or more usually by its central or reserve bank, on behalf of government. However, the stock of money is in excess of the amount printed since other pieces of paper count as money or 'quasi-money'. In effect, when you buy goods by a cheque (guaranteed by a commercial bank) the bank, in honouring the cheque, is creating money, for the seller can obtain money for your cheque on presenting it, through *his* commercial bank, even though your account may be overdrawn, i.e. even though you have insufficient money on deposit in your account. It is not actually the cheque which is money, but the cheque deposit. The creation of such money has been achieved by the commercial bank which has granted its client cashing facilities. In our sophisticated society it is possible to live without handling money by giving and receiving cheques; the credit card is, in effect, merely a means of writing cheques monthly.

There are conventions (usually controlled by law) as to the amount of cash (or deposits with the central or reserve bank which count as cash) which a bank must hold against its cheque deposits. Let us call this r, and assume that this is 10% and that bank A has a client who deposits $1000, ($D$).

Bank A now has $900 it can use, and will operate as follows:

Assets ($)		Liabilities ($)	
Reserves,	100	Deposit,	1000
Loans and investments,	900		

In other words it will make loans (or overdrafts) and invest $900 thereby. However, the person who has received the $900 loans will deposit this in another bank, bank B, whose balance sheet will become:

Assets ($)		Liabilities ($)	
Reserves,	90	Deposit,	900
Loans and investments,	810		

and so on to banks C, D, E, etc.

Eventually the total money in circulation will be $900 + $810 + \cdots = $10\,000 = D/r$. The consolidated balance sheet of all banks combined will now be:

Assets ($)		Liabilities ($)	
Reserves,	1000	Deposits,	10 000
Loans and investments,	9000		

Thus the original deposit of $1000 has become $1000/r$ (in this case $10 000). *Similarly it follows that, when a deposit is withdrawn, a multiple reduction in money supply takes place.*

9.4   THE BANK'S MULTIPLIER

If we call the bank's reserves R_b, and the deposit D, then $D/R_b = 1/r$ when r is the legal (or actual) ratio of deposit held in reserve. Then $1/r$ is called the bank's multiplier. However, in practice the multiplier is smaller than this as the client (the private sector) will retain a small amount of *cash* for immediate demand needs, and the multiplier will reduce similarly. If the client retains $q\%$ of his funds in cash, the multiplier becomes not $1/r$ but $1/(r + q)$.

Note, then, that the bank's multiplier is largely controlled by the Central Bank policy in defining the value of r, and to a lesser extent by the private sector in deciding the amount of reserves it too wishes to retain.

Just as there is a bank multiplier, there is a money-destroying multiplier when the central bank withdraws money from the commercial banks; it can do so by calling in money from them by the sale of bills and by other means, increasing the commercial bank's demand deposits with the central bank.

9.5   THE CIRCULATION OF MONEY

Of course, the private sector, in arranging for an overdraft (which is a commercial bank loan), will have to pay an interest on the loan to the bank. Indeed, although these banks may charge their clients for cheques and for statements, a major source of income to pay their expenses and profits is the interest charged on such loans. Thus, if the interest is high the private sector will borrow less, and the bank's actual multiplier will be less than its theoretical multiplier.

9.6   THE BANK RATE

This is the legal rate laid down by the central bank for money borrowed from it by the commercial banks. Thus, when a commercial bank obtains an income by giving loans to its clients, it will charge the bank rate, plus a handling fee. Such a handling fee will vary from a small amount — perhaps 1% for a big loan to a reputable company with good security — to a big percentage for small sums to smaller firms or individuals where the risk is higher. In this manner a market loan rate is derived.

In all cases a rise in the bank rate will raise the market loan rate and will cost the commercial bank's clients more, and they will thus voluntarily tend to reduce their borrowings, thereby reducing the amount of money in circulation. Indeed a per cent or two rise may make some potential schemes unviable. This reduces the money available for engineering schemes and works.

9.7   THE VELOCITY OF MONEY

The amount of money in circulation is defined by the so called *velocity* or money equation,

$$MV = PT$$

where M is the amount of money, V the average velocity of circulation (the number of times each dollar is spent), P the price level (the average price per unit sold), and T is the number of transactions (units of goods sold), all in a given time period. On the macro scale, if T is regarded as the real goods and services provided in a year, and P as a gross national product deflator (a correcting factor for inflation), the $MV = PT = \text{GNP}$; but this is a special case.

From the velocity formula, $MV = PT$, it is apparent that if V and T remain equal, a reduction in money in circulation (say due to the manoeuvres of the central bank) will reduce prices, and conversely an increase will raise prices. For example, when trade unions force a doubling of wages, more money has to be created to pay them, and in the long run prices will increase. The exception to this would be the case when any increase in rates is coupled to an equal increase in productivity, when prices will remain constant, and the extra money, M, is circulaiton would tend to cause the number of transaction, T, to increase, and more goods would be available.

Of course, as M is reduced there could be a tendency for V to increase, or for T to decrease. If T decreases, then less goods and services will be produced, and the signal is given for unemployment to increase. Thus the control of the money supply is a difficult art and requires balancing the evils of inflation and unemployment.

Chapter 10

Taxation

10.1 TAXATION AND MONEY SUPPLY

Taxes are raised by governments; firstly to provide revenue to finance the budget for the civil service, for such services as Public Works, Defence and Security, etc.; and secondly as a further control on the money, M, in circulation. If a government were to reduce taxes (or print and issue more money), the increase in money supply in the private sector, other things being equal, could cause both more inflation (an increase in P which is a reduction in the purchasing power of money) and more transactions, T, which would give more employment.

10.2 THE ACCELERATOR

An increase in employment will result in further increases in demand and hence a further increase in transactions, the rising consumer spending resulting in rising sales. Rising sales will deplete stocks (inventories, in economic terms) and firms will increase production. Sooner or later, the increased level of production will call for more investment (for which money is available) even if only to replace obsolete plant. This is known as the *acceleration principle* or simply as the *accelerator*.

The accelerator thus starts the chain reaction of more investment, more income, more consumption, more investment, etc. If consumer expenditure continues, there will be an economic upsurge; but if the extra income generated is not spent but saved (not to be confused with investment in economic terms) the acceleration will stop, and only a hump in the economy will result. Like the multiplier, the accelerator can work in reverse, and reduced investment can start a chain reaction resulting in a depression.

10.3 GOVERNMENT EXPENDITURE

We have shown how a reduction in taxation, by making more money available in the private sector, should revitalize the economy; but it must be remembered that with less taxation income the government itself will have less to spend, and must curtail expenditure; this has two reverse effects.

Firstly, with the revitalized economy and extra spending in the private sector, inflation is likely since P will tend to rise with T, and there will be a demand by civil servants and other governmental employees for increased salaries to meet the cost of living, yet the government's income will be smaller. Secondly, and more importantly, the government will have less money available for its expenditure on and expansion of such services as Defence, Education, and Public Works, and may have to reduce its commitment to these areas. This reduced expenditure will thus act as a deflator, acting against the accelerator of the private sector.

The solution, in this case, may be for government to accept a deficit in its budget, that is to budget 'in the red', or to commit itself to spending more than it receives in taxation. For a year or two government may continue to do this (it will obviously never be able to budget *exactly* from year to year) and indeed it may have reserves from previous years, but patently it cannot continue with deficit financing for a long period as this will have other effects, not the least of which will be its international consequences.

The use of taxation, or of government borrowing and expenditure, to control the economy is known as the *fiscal policy*, whereas the control of the amount of money by the central bank through the commercial banks, and by the commercial banks themselves, is known as *monetary policy*.

It should, perhaps, be noted that the fundamental problem in the implementation of both policies is that of timing, since it is necessary to apply one on upturns and the other on downturns of the economy, and it is not always known at which part of the business cycle the economy may be at any given time.

The famous economist Lord Keynes was the originator of much modern thought on the subject of monetary and fiscal control, on the interest rate, and of the need for deficit spending as a means of controlling depressions.

10.4   TYPES OF TAXATION

Taxation is classified as *direct* or *indirect*. Direct taxes are those levied on income (income tax) and on death duties (again levied on income), etc. Indirect taxes are customs duties, sales tax, value added tax and the like. Taxes are further classified as *proportional* (a fixed percentage of price, such as a sales tax), *regressive* (such as a poll tax per head of population which, percentage-wise, is smaller with increased income), and *progressive* (such as most income tax systems which take higher percentages of larger incomes).

Government adjusts total taxation to give an income satisfactory to its needs, and varies the nature of taxation from regressive, through proportional, to progressive, according to its social, political and economic goals; it then redistributes some of its income in the form of subsidies which are, in effect, negative taxation.

For example, it may decide to subsidize bread, corn, milk or other national staple diets, so that, irrespective of the cost of production, every citizen is assured of a minimum of necessary calories. Of more interest to engineers is the usually hidden subsidization of the transport infrastructure. Government in under-developed countries usually subsidizes commercial road transport by spending more on roads than it receives in vehicle licences and fuel tax; it may provide railways with a free right of way, or it may subsidize municipalities with contributions to urban transportation. Again it may provide dams for urban water supply and not pass on full costs to municipalities that draw therefrom.

10.5   THE EFFECTS OF PROGRESSIVE TAXATION

There are four important effects of progressive income tax.

The first may be described as a social effect. For example, imagine that society makes a subjective judgement that a managing director is five times as productive as a labourer, and that he should have five times the salary that the labourer receives. If tax is proportional, the director will have five times the labourer's income, after tax. However, if taxation is progressive, as is usual, he will end up with less than five times the labourer's income, which defeats the initial goal.

The second is the equalizing effect of progressive taxation in inflationary times. If, say, over a period of time, inflation sets in and wages increase. Everyone is then in a higher income group and pays a higher percentage of his income in taxation. Thus, in inflationary times, government is better off, and the private citizen worse off, always assuming that goods have increased in price as much as have wages. There will also be a change in the ratio of after-tax spending power between the manager and the labourer, which may well mitigate even more against the higher paid manager.

The third and most important aspect of progressive taxation — which explains why it is followed by most governments — is its stabilization effect on the business cycle, and particularly on inflation.

The mechanism is as follows. With increasing inflation, prices increase and wages, possibly due to cost-of-living allowances or trade union negotiations, also increase. Since the taxation is progressive, the workers now pay higher percentages of their increases in taxation, the higher the income, the higher being the percentage. Thus, proportionally, the now better paid workers have less to spend than before, compared to the increased prices of goods. However, the government receives proportionately more in taxation, and by the withdrawal of proportionately more money (or spending power) from the private sector, provided that the government curtails its own spending, a stabilizing brake is put upon the inflationary cycle. Remember $MV = PT$; with less money in circulation, prices should tend to decrease.

The final aspect of progressive taxation is that the incentive to work harder or longer, is destroyed since an increasing percentage of earnings is withdrawn as taxation. This explains the present tendency to increase proportional taxes, such as sales taxes, and to reduce income tax. An increase in fuel tax — another indirect tax — could well be a better alternative than to raise progressive income tax. In this case the motorist who wishes to travel excessively would pay more and would have an encouragement to work harder to pay for this privilege.

Chapter 11

Trade and Economic Growth

11.1 ECONOMIC GROWTH

Economic growth is associated with the present century and follows from improved technology (both in agriculture and industry), from improved productivity (both from improvements in technology and the acquisition of skills by training), and from improved transportation and foreign trade. Whilst extractive industries such as mining and farming are classified as primary, and construction and manufacturing as secondary, transportation, like government services, is classified as a tertiary industry. In a highly industrialized society tertiary industries can employ two thirds of the labour force, whilst primary industries can employ two thirds in underdeveloped countries.

Transportation improvements assist international trade. Some countries, either due to their favourable technological skills, soils, climate, or minerals, are able to produce certain goods more efficiently than others; this provides an incentive to improve transportation, and thus trade, between countries, which in turn increases economic growth in *all* the countries concerned.

As economic growth occurs — and its association with productivity and technology can readily be seen — national income increases and with it the standard of living of the citizens.

Increased prices are normally associated with growth in national income, and in determining real growth it is necessary to refer back to a datum, such as 1900 prices. However, as long as national income increases faster than prices, real economic wealth is increasing. Thus, although national income of most countries may have been lower in the Great Depression of the thirties than it was in World War I, it was probably higher then than it was in Victorian times. That is to say, the average citizen was more productive and better off in 1930 than in 1830. It is generally accepted that maximum economic growth is associated with creeping, as opposed to galloping, inflation.

As we have seen, one reason for economic growth is international trade, and although detailed calculations are irrelevant to an engineer, an understanding of the *theory of comparative advantage* is worthwhile.

11.2   THE THEORY OF COMPARATIVE
ADVANTAGE

Since the balance of payments and international rates of exchange have not yet been discussed, it is convenient to consider production of goods in terms of labour output. Indeed, productivity is real, and money merely an artificial means of measuring productivity in labour and materials.

We know that the USA and Japan are about equally efficient in the production of cars, but that the USA is more efficient in producing wheat. Let Table 11.1 represent the production of each country per man-year.

Table 11.1   Postulated production of USA and Japan

Comparative production per man-year		
	USA	Japan
Wheat (bushels)	200	100
Cars	1	1

The theory of comparative advantage states that both countries will improve their standards of living by specializing in the production of the goods in which they have *comparative* advantage (provided that transport costs, which are ignored, do not negate this advantage).

Consider a group of a hundred workers in each country, and suppose that fifty of these worked on the production of each product; then yearly outputs, before international trade and specialization, would be

USA	50 workers . . . . 10 000 bushels of wheat
	50 workers . . . . 50 cars
Japan	50 workers . . . . 5000 bushels of wheat
	50 workers . . . . 50 cars

Total 'world' production = 15 000 bushels of wheat + 100 cars

Suppose now that the US workers only produced wheat, and the Japanese produced only cars. After such specialization we get

USA	100 workers on wheat	20 000 bushels
Japan	100 workers on cars	100 cars

Total 'world' production = 20 000 bushels of wheat + 100 cars

Thus with the same labour input, total production has increased by 5000 bushels of wheat. Indeed, even if Japan were a little less efficient than the USA at car production we could still increase 'world' production by specialization. It follows that with international trade, all workers could have enhanced standards of living. How much of such trade should be undertaken is a separate study;

suffice it to say that if total world production can be increased, trade will follow and both countries will gain.

The extent of such international trade must obviously be limited by transport costs and by the practical limits of specialization, either those of worker potential or of land (in the economic sense) availability, with money (in our economic society) being available to replace barter.

Digression. Western countries were often accused of raising their standards of living by exploitation during their colonial expansion. However, it can be seen that this did not necessarily follow; by developing a colony both it, and the colonial power, could improve their standards of living through the working of the theory of comparative advantage. Similarly it follows that once the colonies had been developed and trade patterns established to mutual advantage, there remained no economic case for the maintenance of the colonial system.

11.3 THE EFFECTS OF CUSTOMS AND TARIFFS

Customs and tariffs are imposed by governments to provide a source of revenue, and to protect local, often less efficient, industry. Customs duty and tariffs are similar to increased transport costs; the enhanced artificial prices hamper the free flow and the balance of trade, thus reducing the total maximum efficient production level, which in turn reduces the standards of living in both *importing and exporting* countries. In general the lower customs dues are, the better it is for all concerned, and this principle underlies the formation and policy of the European Economic Community (the Common Market).

In the real world there is an infinite variety of countries, and of goods, produced. Each country tends to specialize and export goods which it can produce *relatively* efficiently, with limitations due to the amount of land and labour available for specialization, and the use of money replaces the need for direct barter. Taste will also play a role, for the 200 workers in the example may wish not to have 20 000 bushels of wheat and 100 cars, but say 25 000 bushels of wheat and less than 100 cars. Indeed, they may wish to work less hard and have less of each, in which case, by trade, they can have the same amount of goods as before, but by working less hard.

11.4 INTERNATIONAL BALANCE OF PAYMENTS

To finance trade, money is used as a medium of exchange. However, if a small country, such as Zambia, is trading with another, such as Sweden, and wishes to import Swedish cars, it is unlikely to have Swedish kronen available; similarly Sweden is unlikely to be willing to accept payment in kwachas, particularly as it may wish to buy no Zambian goods. Thus one of the major international xeno-currencies, such as US dollars, or deutschemarks, or Swiss francs, or Japanese yen, will be transferred from Zambia for the Swedish car purchase. Of course, gold or 'paper gold' of the International Monetary Funds, could also be used instead of an international currency. In all cases the money is transferred from the holdings of the central bank of the importing country, which replenishes its holdings of gold, US dollars etc., from reverse trade, which need not be with the exporting country.

It follows that if a country exports more than it imports, it will build up a reserve of foreign currency (or gold) in its central bank; but if the reverse is the case, it will deplete its reserves. Thus a country cannot operate indefinitely with less than equal terms of trade (the ratio of export to import value). Exceptions to this are when the exporting country is willing to invest in, or lend its money to, the importing country. This has been the case with Arab oil money in the UK.

11.5   EXCHANGE RATES

Another problem is that of establishing an exchange rate between, in the example, kwacha and kronen. This may be achieved by each country establishing the value of its currency in terms of such currencies as the US dollar; the ratio held of its money in circulation to its reserves of gold, or of important foreign currencies such as the US dollar, establishes the credit worthiness of that currency. The problems of the gold standard, or fixed or floating currencies, need not concern the engineer, but those of devaluation or of revaluation do.

If, over a long term, a country continues to import more than it exports (in terms of gold or of a foreign xenocurrency) it must establish import controls (to discourage imports), or raise tariffs on imports (for the same reason), or devalue its currency, which by raising the local price of imported goods, discourages these imports. However, in raising tariffs, or in imposing import controls, the country is no longer engaging in free trade, and will thus cease to gain all the advantages of free trade which follow from the theory of comparative advantage.

11.6   DEVALUATION

Suppose a country devalues its currency. By doing so it will reduce its total imports since individual imported items will cost more in terms of its local currency, and increase its exports, since they will cost the same in local currency but less in foreign currency to the buyer. The total amount of foreign currency acquired will depend on both the demand elasticity of the importing country and the supply elasticity of the exporting country, since more goods must be exported at the reduced price in terms of foreign currency to earn even as much, let alone more, foreign currency as before. As a result, even in the short run, it does not follow that devaluation will result in the acquisition of more, thereby reducing the deficit of, foreign currency. Indeed, if the demand is inelastic the position will worsen, as the same amount of exports will result in less foreign currency inflow.

In the long run, the higher import costs (in terms of local currency) will increase total costs of production, which may trigger off inflation and cause wages to increase. In turn increased wages will raise the cost of production of goods, both for internal use and for export; this will reduce the demand for exports, and hence the revenue earned by them, so devaluation cannot be regarded as a long-term panacea for correcting balance-of-payments problems. For this reason, many countries have import controls (restrictions on the use of foreign currency) and simultaneously try to increase productivity, which is the only certain cure for economic ills.

11.7   INVISIBLE EARNINGS

Whilst these are of little direct interest to the engineer, it is advisable for him to know of their existence. A nation can earn foreign currency not only by the export of goods, but also by the export of services. For example, Liberian ships may carry Japanese goods to the USA, and their voyages may be insured by Lloyds of London, thereby earning foreign currency. British consultants may supervise German contractors in Iran, and holiday-makers also generate foreign currency in the country they visit.

The investment of capital by a developed country in another country earns interest and thus generates foreign currency. Many countries have achieved favourable balance of payments by such invisible earnings, even though their physical, or visible exports, are negligible.

Chapter 12

The Costing of Foreign Currency

12.1 THE CONSERVATION OF FOREIGN CURRENCY

No country is fully self-sufficient in its production resources, and certain minerals and machinery have to be imported. If a country is lucky enough to have abundant minerals to trade such as petroleum crude, or foodstuffs such as wheat, it will receive foreign currency in return, and can use this freely to import other goods which it needs. Indeed, for example, if Saudi Arabia needs cars it would be ludicrous not to spend foreign currency and import them, since it cannot produce them more cheaply than it can import them.

However, if a country has few wanted exports, its stock of foreign currency will be limited and must be rationed for essential imports. Indeed, if any country spends its stock of foreign currency, two effects must be considered. Firstly, in depleting its foreign currency reserves it cannot import foreign goods unless foreign countries will grant loans or invest in that country. Secondly, in spending money outside of its boundaries, the country is withdrawing money from internal circulation. The formula $MV = PT$ now applies. If M is reduced by expenditure overseas, the effect is the same as increasing the commercial banks' holdings with the reserve bank, and there will be a tendency for the number of transactions, T, to reduce. In turn, production will fall, and unemployment may be triggered off. The unemployed workers will themselves consume less, create still more unemployment, and the process will continue; in other words the multiplier and accelerator will work in reverse.

It follows that whilst the increase of foreign currency in a country will create employment, its reduction will reduce employment; governments must thus conserve their stocks of foreign currency.

12.2 THE CASE FOR SHADOW PRICES

The problem facing all countries to a greater or lesser extent is not so much how to ration their stock of foreign currency, but when to do so, and how to assess its worth. In the days of the British Empire, empire preference existed; empire governments often reduced tariffs on empire products, and weighted foreign tenders by an arbitrary percentage. Today it is more usual for governments to use the shadow price concept; indeed the World Bank uses these as a basis for development investment.

The shadow price is merely a ratio with which the landed price of imported goods, or services, in local currency is multiplied by the government of the importing country, to assess the value of the item against the loss of foreign currency from reserves of its central bank.

It is important to note that the private entrepreneur, needing to import to make his project viable, is not concerned with shadow prices — he will either be permitted to import or will be refused such permission by his bank, which follows government directives. The shadow price concept is, or can be, used as a tool by government in assessing the viability of its own, or of the private sector's, projects. It can be said perhaps that the entrepreneur assesses the financial viability, while the government assesses the economic viability of projects.

Indeed, for the engineer and economist working in, or for, underdeveloped countries in particular, shadow prices can be all important in establishing the economic viability of a scheme. Examination of published lists of unofficial exchange rates for free and uncontrolled buying and selling of currency which exists in so-called 'free exchange markets' in such places as Hong Kong, will give *some idea* of possible shadow prices, but not a true figure for certain national currencies circulating there cannot be legally repatriated without a permit from the country's central bank.

Shadow prices may occur through four reasons:

(1) *Official exchange rates.* These, with import controls applying, may not reflect the scarcity value of foreign exchange. The shadow price of foreign currency will thus be higher than the official price, often several times.

(2) *Minimum wage rates.* These, if established by law or by trade unions, may not measure the true costs or worth of the productivity of labour. Real costs are certainly low if underemployment exists as it may be better to create work than to import goods. Skilled labour might be 25% higher in value, and unskilled labour 50% less than controlled wage rates, in terms of real costs, that is in terms of productivity.

(3) *The real or opportunity cost of capital.* This is measured as the interest rate. In many developing countries it may be 10% or even 20%, due to risk; yet foreign government funds may be offered at far lower interest rates for political reasons. Such reasons could be goodwill, or insistence on spending the money in the country giving the loan. This results in schemes which could be viable at low, but non-viable at real or commercial, interest rates.

(4) *Taxes.* Taxes affect the shadow price, i.e. indirect fuel taxes and import tariffs affect the *financial* costs, but should not be included in economic cost analysis without an adjustment for shadow prices. It is apparent that a project in the private sector has to allow for these taxes and tariffs, whereas a similar government project does not need to do so.

12.3   THE USE OF SHADOW PRICES

By definition the shadow price is merely the ratio by which the price of imported goods, or services, in local currency, is multiplied. For example, if our currency is the franc, there are 4 francs to a US dollar at the official rate, and we have the choice of a locally made motor at 20 000 francs, or an imported one (ignoring duty) at US $4000 (that is 16 000 francs), which would we buy? Consultants to a firm would recommend the imported motor as it is cheaper, but if they were consultants to the state the problem would be different.

Using the shadow price concept we must multiply the price of the imported item by the shadow price. If a decision has been made that this is, say, 1.5, then the imported motor is $16 000 \times 1.5 = 24 000$ francs in adjusted cost to the state, and the state would buy the local motor at 20 000 francs. However, were the shadow price only 1.2, the imported article would be valued at 19 200 francs and would be selected.

The determination of the shadow price is the prerogative of the state. If the balance of payments were extremely adverse, a high shadow price would be adopted to ration foreign currency; whereas if it were favourable over a long period, the shadow price would be unity.

Of course, the shadow price is abstract, and consulting engineers preparing reports for government may not be advised of its value. In this case their appraisals must be open-ended. The price of the local motor was 20 000 francs, and that of the imported one 16 000 francs. They should say that if the shadow price exceeds 1.25(20 000/16 000) the state should buy locally, and vice versa, and leave it to the Treasury or central bank officials to decide on the action to take.

Of course, the abstract shadow price will vary continually and the engineers' government clients may not themselves be aware of a figure, or ratio, to adopt. In this case the consultants can work out the various rates of return, or cost–benefit ratios, on different solutions to a problem; the scheme which is viable with the highest shadow price is the better scheme to adopt.

Few schemes are constructed with solely foreign currency expenditure; for example, a hydro-electric scheme, at least in a developing country, is likely to entail foreign currency for its electrical and mechanical components, and local currency for much of its civil engineering content. The former costs are multiplied by the shadow price and added to the latter to arrive at the *economic cost*.

Finally it must be noted that shadow prices also work with exports. If the hydro-electric scheme were to make aluminium for export, the f.o.b. value of the aluminium should be multiplied by the shadow price. Indeed, in certain circumstances there could be economic justification for running a state enterprise at a loss, or even for the state to subsidize private enterprise running at a financial, or 'book' loss, if it is exporting its products when foreign currency is in short supply.

Example 12.1

In a country with a shadow price of 2 a private irrigation scheme is planned, which will cost $100 000 p.a. in local, and $50 000 p.a. in foreign currency. The production can be exported for $120 000 p.a. Should the state give a subsidy?

The entrepreneur's annual financial balance sheet can be written

Expenditure	Income	Loss
$150 000	$120 000	$30 000

However, as a state project, whilst the financial balance would be the same, the economic balance sheet would be

Expenditure	Income	Profit
$100 000 + 2 × $50 000	2 × $120 000	
$200 000	$240 000	$40 000

The entrepreneur would close the scheme, but it is worth $40 000 p.a. to the state. Thus a subsidy of not less than $30 000 would make the private enterprise scheme financially viable, and any subsidy not exceeding $40 000 p.a. would be in the state's interest.

It is worthy of note that if the $50 000 in foreign currency were for petroleum, and $50 000 of the $100 000 local currency were a fuel tax, the cost to private enterprise would have been unaltered, but the *financial* cost to the state would have reduced to $100 000 from $150 000 as the state is unlikely to charge itself taxes. Thus the state project would show a *financial* profit of $20 000 p.a.

PROBLEMS

Problem P12.1

A deviation is proposed on a state-owned railway which will cost $1 000 000 and will involve only local currency. The ruling interest rate is 5%, and the shadow price for foreign currency is 3. The scheme involves shortening, on existing level track, by 1 km. The operational cost of trains is $10 per kilometre, half of which is in imported fuel. How many train movements per year are necessary to justify the construction of the deviation, that is to say when operational savings are equal to or greater than operational costs?

Problem P12.2

If the railway in Problem P12.1 were privately owned, and 3000 train movements per year were involved, should the owners proceed with the scheme?

Problem P12.3

If the railway in Problem P12.1 were privately owned and used by 4000 trains a year, would the state be justified in granting an annual subsidy, and, if so, how much?

Chapter 13

Public or Welfare Economics

13.1 INVOLVEMENT OF ENGINEERS IN WELFARE ECONOMICS

The state has to perform the economic roles of providing public goods, such as the transport infrastructure, and of controlling externalities, such as the external nuisance of pollution caused by a factory. Since many engineers are employed on public sector works, and since most engineering projects involve externalities, some understanding of welfare economics is essential.

13.2 PUBLIC GOODS

These differ from private goods in that they are usually indivisible, and are shared by all citizens to a greater or lesser degree, irrespective of the amount of tax the individual may pay. Public goods are normally regarded as indivisible in that the costs cannot be allocated proportionately to the use. For example, we all use roads but we cannot pay our share according to our use of them.

An electricity undertaking (often municipal, or quasi-government) can allocate its costs directly to the user, but no foolproof scheme can be devised whereby each user pays his due share according either to its cost, or to its value to him. This explains the electricity tariff system. The small private user in a village may, for example, be prepared to pay 2 cents, when average costs would be 3 cents per kilowatt hour. If, however, a large industry is established willing to pay 1.5 cents, the economics of scale may reduce costs to 1.7 cents; the scheme then becomes viable if the private user pays 2 cents, and the industry pays 1.5 cents per unit. This is usually achieved by a tariff system with a high price for the first few units consumed and a lower price for subsequent ones. Here the user does not pay for his true share of the costs, but the differential tariff makes the supply of electricity feasible to all.

Road costs for construction and maintenance in developed countries are often more than covered by vehicle licensing and fuel tax, though the commercial vehicle operator seldom pays for his share of the damage he does. Indeed, it is difficult to assess accurately fair charges for both damage and road occupancy, and also for the delay costs the juggernaut lorry imposes on the private motorists through congestion.

Again, a similar road in a developing country may carry only 10% of the traffic but will cost 50% as much, and taxation income will be in adequate to finance it. However, the road may be necessary to enable the country to develop. Moreover, its construction would cause land values to rise. In such cases, since fuel and vehicle tax cannot equal road expenditure, the road is normally financed, directly or indirectly, from general taxation (or from rates in a municipality), and costs are not apportioned according to individual use.

13.3   FUNDING OF PUBLIC SCHEMES

Public schemes are usually paid for by taxation, which goes to the revenue account; this is used either directly to finance them, or indirectly to service loan accounts. A loan account, as its name implies, is funded by borrowing and is repaid, with interest, from subsequent revenue. In developing countries it is usual to finance public works from loan accounts; the construction scheme enables national income to grow from the use of the works which, in turn, makes it possible to repay the loan. Contractor finance is similar in principle, but not so desirable since the client can seldom obtain bids competitive in both unit rates and interest charges.

In developed countries construction is often funded direct from the revenue account as the standard of productivity is already high enough to enable future investments to be paid from current earnings.

13.4   EXTERNALITIES

This is the name given to external effects, either social or economic, resulting from economic actions, which are almost invariably actions of engineers. External economies are social gains, external diseconomies are social losses. Both types of externalities may result from the establishment of a factory. The manufacturing process may result in social diseconomies, difficult to quantify but nevertheless real, in terms of air, water, and noise pollution, which may cause neighbourhood land values to drop. A further diseconomy may be the need to provide extra roads for the traffic generated by the workers. Conversely an external economy may be an increased demand for housing which could raise certain land prices.

The construction of an international airport will certainly have externalities; diseconomies will result and property values will drop under the approach funnel, whereas economies will result with increasing land values in other areas nearby. Similar results occur when a suburban railway is built; adjacent land values fall, nearby land values rise. In developing countries, external economies tend to outweigh diseconomies. Often large mines are established off the beaten track. In such cases these mines may require to finance, or to supply finance towards, the construction of a concomitant infrastructure. A drive-in cinema can be made to pay for a new road and be exempted, for a time, from entertainment tax. Land alongside the road or rail to a new mine has immediate access thereto, and hence to a market. The land may then be used for dairy purposes, instead of ranching, to supply the new market. This is classified as a social economy. Conversely a high voltage power line, or a pipe line, creates diseconomies.

Cost–benefit analyses are comparisons of the economic costs of schemes with their economic benefits, and when preparing analyses in such circumstances, the engineer should allow for this increased land value as an external economy, i.e. he should allow for this external benefit. It may not benefit his client directly, but it will benefit the state. For example, if the output of a new mine could be taken to market at equal cost to the mine owner either by road or by an overhead rope conveyor (the latter only having external diseconomies) the owner would, *ceteris paribus*, obviously find the road scheme more viable; and moreover, he could ask for state assistance towards road costs.

Perhaps the mine also pollutes the natural streams, in which case the police roll of government would be introduced. Pollution is an external diseconomy, which, assuming that it cannot be tolerated, can be assessed economically. A scheme must be established to rectify the pollution; the producer could be required to do this by law (thereby reducing his profits), or the government could undertake the work, taxing the producer for its task. If the government is weak and allows the pollution, the external diseconomy will continue, downstream farmers and fishermen will suffer and might even demand compensation. This diseconomy must still be included in cost–benefit analysis.

PROBLEM
Problem P13.1

An entrepreneur wishes to develop a long-life mine for $1 000 000, and the market interest rate is 6%. His operating expenses are estimated at $100 000 p.a. Nearby landowners demand $100 000 compensation for air and dust pollution, but the scheme includes a road costing $300 000 which will add $200 000 to land values of property adjacent to the road. Determine the annual profit of the mine, if estimated sales are $200 000 p.a.

Chapter 14

Inflation

14.1 DEFINITION OF INFLATION

There are many definitions, but perhaps the best for the engineer is that inflation is the rise in price level of currently produced goods and services, or a rise in the GNP *per se*. Just as there are many definitions there are equally many theories and explanations of its causes and effects. However, no one will doubt that it will occur if productivity falls.

The engineer is used to thinking and planning in terms of productivity and output. To him money should be merely a convenient measure of these, and in inflationary conditions it is usually money, not productivity, which changes. If prices rise at the same rate as wages and materials, as well as incomes from investments, rents, pensions, etc., and productivity and employment levels remain unchanged, there is little to fear. Unfortunately pensions tend to remain static; indeed it has been said that inflation causes no harm save to pensioners and others who deserve well of their country!

14.2 METHODS OF CONTROL OF INFLATION

This control is a difficult art. In view of the velocity formula of money, $MV = PT$, the withdrawal of money from circulation should cause prices, P, to drop and thus stop inflation. However, it also tends to reduce the number of transactions, T, with the result that less goods are demanded, and unemployment increases.

Of course, in inflationary periods people tend to purchase, rather than save deflating money. This also raises M, and hence P, so governments raise interest rates to attract this money into government bonds and stock, not because they need it, but to withdraw it from circulation. This raises the interest rates demanded for private sector investments, which cause the values of shares in existing projects to drop, and hence reduces the wealth of investors and the viability of planned projects. The number of projects commissioned thus decreases, and again unemployment is triggered off.

If governments of large countries have difficulty in controlling inflation using such monetary policy, then governments of smaller countries have even greater problems. Whilst they could apply similar measures to acheive success, they will be forced to 'import' inflation from larger countries with whom they trade.

14.3   THE TIMING OF PUBLIC WORKS

The effect of inflation is vital to engineers. They are often faced with apparently plausible advertisements in trade journals, recommending the advancement of contracts, particularly those for preventative maintenance, when inflation is rampant; these are based on the specious claim that the work will cost more if it is postponed. This reasoning is quite fallacious for the cost of doing a job today (if technology remains unchanged) will be the same in terms of labour, or the current dollar, whether it is done today or tomorrow even though work postponed until tomorrow will cost more in terms of tomorrow's dollar, which will be worth less than today's. The engineer, faced with such problems, must ask the following questions before making a decision.

(a) Will technology or productivity improve if the scheme is postponed? If so, costs at *today's dollar* price will drop if it is postponed.

(b) Will materials increase in price more rapidly than labour? If not the scheme might be postponed. This will depend upon the elasticity of supply. For example, the world is rapidly depleting its mineral reserves, especially of petroleum crude from which bitumen is derived. Thus it is reasonable to expect that bitumen, in inelastic supply, will rise in price faster than will labour, so perhaps the work should not be postponed but undertaken straight away. However, if maintenance is regular, say resealing a road every five years, the advancing of one operation is a commitment to the advancing of all future operations, which adds the complication of interest rates and discounting.

(c) What is the cost–benefit ratio? If the scheme will produce large profits, it should be proceeded with as soon as possible.

(d) What is the current position on the business cycle? If the scheme is a long-term one, like a large dam, it will consume investment for several years before earning income, the implications of which must be considered. Contract tenders usually include price escalation, but not reduction, clauses. Thus it is not advisable to award long-term contracts at a peak in the cycle though it is a good time to contract for them.

Conversely, if one is at the trough in the cycle, i.e. a recession, prices will probably be lower initially than they are likely to be at any future date, so it is a good time to award such contracts. Also on a national (or macro) scale, the injection of money into the economy at troughs by awarding such contracts will help to trigger off the recovery.

Under this heading it would perhaps be wise to remember that large engineering works, such as motor road programmes, are undertaken by governments, not only to meet present or future needs, but also as a means of reflating or controlling the economy, the price level, and employment. Many motor roads were built in South Africa before traffic justified their construction to create employment and thereby to maintain the increase in national income.

14.4 INFLATION AND CAPITAL INTENSIVE WORKS

In the transportation field, schemes such as pipe lines are often advocated in inflationary times, since they are capital intensive. Their proponents claim that as initial fixed costs are high, only the small operational costs will inflate, so that future transport costs can be contained. They maintain that with inflation wages will increase, and in the future we will pay proportionately less of our income for the transport.

The argument can be valid only if the service is to be short term, and will not need replacing on being written-off. To simplify this argument, let us assume that inflating operational costs are so small a part of the total cost that they can be neglected, and only the initial capital for a pipe line needs to be considered. The charges for transport, *if based on original cost,* will be low, and if the worn out pipe needs no replacement the entrepreneur, at fixed transport prices, will get back his capital and profit. However, this capital will be in inflated dollars, worth less than the original investment in buying power. Should the service be long term it would be inadequate to replace the pipe line at the inflated prices then ruling. Thus we can conclude that each day, month, or year — whatever unit of time is our accounting period — the entrepreneur should raise the transport charges so that when the pipe line needs replacement at inflated prices, he has sufficient inflated dollars to do so. Thus even on capital intensive schemes, charges should be raised with inflation.

There is an alternative proof. Suppose we build a pipe line in year 1, and demand is doubled in year 2, but with inflation in prices. Suppose then that we build a parallel, equally efficient pipe line along side the original one; since it will cost more, in current dollars, our new transport prices must be higher. Yet, since it is equally efficient, by definition, its current operating costs must be the same. It thus follows that costs and charges on the earlier scheme must increase to match those of subsequent schemes.

14.5 CURRENT PURCHASING POWER

The modern accounting concepts of *current purchasing power* and *current cost accounting* provide the understanding and tools for the analysis of engineering projects in inflationary periods.

If we take a calendar year as the accounting period, and we buy plant for $\$P$ which will last y years, we can write down depreciation, which must be recovered from operating charges, as $\$P/y$ p.a. However, if there is inflation at $x\%$ p.a. in the first year we will need to replace the plant at $\$P[(100 + x)/100]$ not $\$P$, so we will write our depreciation as $\$P[(100 + x)/100]/y$. Similarly, in the following year it will escalate to $P[(100 + x)/100]^2/y$, and so on. In other words, in writing off plant *it is necessary to forget historical prices*, and to think only in terms of current replacement prices.

It must not be forgotten that the user of a scheme must pay for his privilege, which will include not only his share of current operational costs, which increase yearly with inflation, but also of replacement costs. Thus pipe line transport costs will have high capital costs, and low operating costs, compared to a railway; but, as all costs will be subject to inflation, the cost of using both will increase with inflation.

Example 14.1

A pipe line costing $2 000 000 to build is estimated to last 10 years, and operational costs are $10 000 p.a. Inflation is anticipated to be 12% p.a. effective at the beginning of each year. 100 000 000 l. of oil p.a. are transported, and this demand is likely to last well beyond 10 years. Assuming year-end adjustment of costs and ignoring interest rates, what will be the year-end cost of transporting 1 litre of oil in the first and second years of operation?

At the end of the first year of operation, operational costs will have been $10 000 and 10% of the pipe's life will have been used, replacement costs then being 12% higher at $2 000 000 × 1.12; so the current cost of the year will be

$$\$\frac{1}{10}(2\ 240\ 000) = \$224\ 000$$

Therefore the first year's total costs are

$$\$10\ 000 + \$224\ 000 = \$234\ 000$$

and the cost of transporting 1 litre in the first year is

$$\$\frac{234\ 000}{100\ 000\ 000} = 0.234 \text{ cents}$$

At the end of the second year of operation, the operating costs will have been $10 000 × 1.12 = $11 200. A further 10% of the pipe's life will have been used, but the replacement cost will then be $2 000 000 × 1.12² = $2 509 000, so the current cost of the year's wear will be $250 900. Therefore the second year's operational costs are $11 200 + $250 900 = $262 100. Therefore the cost of transporting 1 litre in the second year is 0.262 cents.

PROBLEMS

Problem P14.1

A bicycle manufacturer requires plant to make 10 000 bicycles a year. The plant must be replaced, with negligible scrap value, every 5 years, and initially costs $100 000. Assume year-end sales and accounting, and inflation at a rate of 10%. Determine the annual cost increase due to inflation to be charged against each bicycle made during each previous year, ignoring any market interest rate and wages.

Problem P14.2

In Problem P14.1 assume: (i) that the materials' costs are estimated at $10 per bicycle, being raised 5% every six months; (ii) that labour is $10 per bicycle, being raised after every twelve months; (iii) that profits, commissions, and overheads are 50% of costs; and (iv) that sales price is constant during the year. Neglect interest rates, and again determine the average sales price of bicycles during the second year of operation if true costs are to be recovered.

Chapter 15

The Financial Operations of Government

15.1 THE CIRCULATION OF FUNDS

Engineers employed in the public sector receive their salaries from government, spend government funds, and pay tax to government. Consultants and contractors form a link in this circle, as indeed does the whole private sector. In the short run government can also spend more than it receives in any given year, so it follows that government's disposable income is its income from taxes or the sale of bonds etc., less any planned surplus, or plus any planned deficit.

15.2 THE ORGANIZATIONAL STRUCTURE

Figure 15.1 gives the simplified outline 'family tree' of a typical governmental organization. The municipal 'tree' is similar, but this time the source of funds is provided by rates, rather than taxes, and subsidies, loans and grants from central government. In developed countries the organization is more complicated, and will vary from country to country; but all tend to be similar in principle.

15.3 CASH FLOW

The central bank acts as a float or stabilizer to the Treasury, just as commercial banks do for individual clients.

Money received by the Treasury is allocated to various government departments as 'votes', so called because the money is approved and voted upon by parliament. Other votes are allocated to the repayment of loans and bonds, contractor finance, and for the payment of interest on other public loans. The control of the amount of money in circulation by the selling or purchase of bonds has been discussed in Chapters 9 and 10.

Money from taxation is allocated to government departments as *revenue votes*. In addition, income from new public loans may be allocated, especially to engineering departments, as *loan votes*; these can only be spent on the construction of new works, such as railways, roads, and dams. Thus loan votes can only be invested in new construction or new capital goods, and never spent on consumer goods or services. For example, it is legitimate for a government to

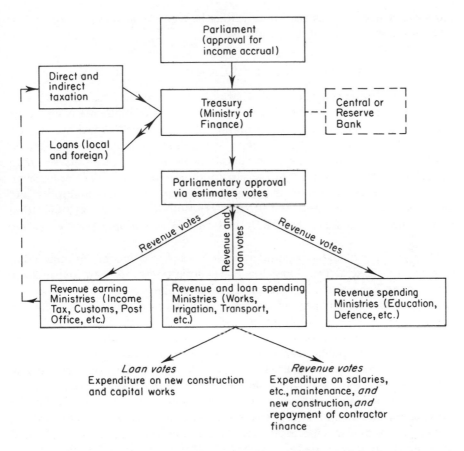

Figure 15.1　Generalized chart of government funding in a developing country

Notes
1. Loan interest is serviced from revenue and loans are liquidated by further loans and by injections from revenue (which increases as a result of development of infrastructure etc. from loan votes).
2. In developed countries most new construction can be paid from revenue votes, due to revenue earning potential of existing infrastructures etc.
3. Commissions and statutory bodies etc. may be self-financing, or may be financed through responsible ministries.

borrow money to build a new dam, or to pay for the nationalization of a factory, but not to pay for its civil servants' salaries, or for its police force.

In turn the departmental votes are split into such items as salaries, allowances, travelling, construction (of specific, government-approved projects), maintenance, repayment of contractor finance, equipment, etc. Ideally, new works should be financed from revenue, but are often built using loan funds, especially in developing countries, where current revenue is insufficiently high to pay for new works. Loan vote-construction is akin to buying a house on a mortgage. The money allocated for construction can either be spent departmentally, or used to pay consultant's fees and for contracts.

15.4 SPECIFICALLY RESTRICTED REVENUE

In some countries specific tax is collected and earmarked for specific expenditure; for example, fuel tax may be allocated for road works. Whilst many engineers press for such schemes, if road expenditure is definitely tied to fuel tax income, then too much, or too little, of the total national revenue may be spent on roads rather than, for example, on education. If there is a recession, less road travel is undertaken, less revenue accrues, contractors are put out of employment, and further recession takes place via the mechanism of the multiplier and accelerator. Treasury cannot allocate such expenditure as part of its stabilization policy, and, on balance, specifically tied income at national level is seldom desirable.

Municipalities often have considerable sources of 'tied' income, as opposed to income from general rates, and this leaves little opportunity for expenditure manoeuvres. Typical of this is the income from parking fees. In many cases this is earmarked for the payment of collection and policing expenses, and for the improvement of parking facilities; it is not available for the provision of other services, such as hospitals, schools or mass transit, which might well be more deserving. Indeed, public transport is essential in any city, but if it is under-patronized it must be subsidized. This may have to come from general rate funding whereas more logically it should be derived from parking fee surpluses.

Similarly in some countries villages are allowed by law to pay for the construction and maintenance of roads from speeding fines. In these cases there is a tendency to impose unnecessary speed limits and to enforce unjustified speed traps merely to finance roads — a policy which brings the local law enforcement forces into disrepute.

Thus it can be concluded that at municipal level there can be no overall justification for specifically tied revenue, and that all income should go to the general revenue account.

15.5 NATIONAL INCOME

Table 15.1 illustrates the computation of national income for a developing country. The details of its compilation are of little concern to the engineer, but he does need to appreciate the need to convert GNI at market prices to those of a base year — 1965 in this case — the difference being due to inflation. Thus, whilst in this table the GNI at market prices shows a steady (if reducing) increase, that at 1965 prices shows an increase followed by a fall after 1974.

15.6 CASH FLOW IN GOVERNMENT

With the best will in the world Treasury experts cannot exactly forecast income and expenditure in the period of half to one and a half years after their study. In

Table 15.1 National income ($ millions)

Year	1965	1966	1967	1968	1969	1970	1971	1972	1973	1974	1975	1976	1977
Wages and salaries	402	412	432	469	513	558	623	694	776	899	1040	1138	1227
Imputed rent	16	16	18	19	22	25	28	31	35	37	39	42	45
Gross operating profit													
Unincorporated enterprises	77	97	127	103	153	139	170	192	163	241	224	247	237
Companies (non-financial)	147	125	150	163	206	240	284	344	418	533	529	536	495
Financial institutions	5	9	11	12	14	16	18	26	30	37	44	46	53
Public corporations (non-financial)	32	18	6	16	22	16	29	36	30	30	42	37	1
Central government enterprises (non-financial)	6	7	6	6	7	4	—	—	—	—	—	—	—
Local government enterprises	11	12	13	13	16	17	18	19	21	19	16	18	20
Less: imputed banking service charges	−11	−13	−14	−16	−18	−21	−23	−30	−33	−42	−52	−56	−66
Gross domestic income (at factor cost)	685	683	749	785	935	994	1147	1312	1440	1754	1882	2008	2012
Plus: indirect taxes	55	50	59	63	69	87	98	107	118	129	137	171	225
Less: subsidies	−1	−2	−3	−2	−1	−3	—	−5	−6	−20	−9	−12	−15
Gross domestic income (at market price)	739	731	805	846	1003	1078	1245	1414	1552	1863	2010	2167	2222
Less: net investment income paid abroad	−26	−19	−13	−15	−18	−21	−30	−35	−39	−40	−38	−49	−44
Gross national income (at market price)	713	712	792	831	985	1057	1215	1379	1513	1823	1972	2118	2178
Gross national income (at 1965 prices)	713	690	751	773	892	924	1023	1111	1155	1273	1250	1190	1105

theory expenditure can be controlled since no ministry may spend more money than it has been allocated; but which authority would not replace a bridge or a dam washed away, simply because the catastrophy had not been forecast?

Although votes are approved annually by parliament, ministers in such cases would authorize emergency expenditure by their ministries, and subsequently seek endorsement of their actions in parliament by means of supplementary estimates.

Forecasting income is even more difficult, and it follows that there will always be a surplus or a deficiency to carry forward to the following year. If the surplus becomes too large, the government can be accused of not developing the economy to its limit, but if a deficiency becomes too large the government, and hence the nation and its currency, will lose its credit worthiness. If this happens either inflation will set in, or a period of underspending to reduce the deficiency will follow, when unemployment — certainly underemployment — of resources will result.

Chapter 16

Project Appraisals

16.1 ECONOMIC ANALYSIS OF ENGINEERING PROJECTS

Engineering works or schemes are not designed or constructed as engineering challenges; in the private sector they are built for motives of profit, and in the public sector, for social benefit, for profit, to minimize the costs of goods, or possibly for political motives. This latter case may be to gain votes, or, more laudably, for an intangible public good, such as improved health or security. Of course, political motivation is not amenable to economic analysis or justification, but it is possible to put a value on social benefits or losses; thus, in all non-political motivation, we can consider the profit motive, which has to be maximized. Project appraisal is the economic study and rating of the costs and values, that is profits, of projects.

To make a profit we need capital to invest and must consider the sources of such capital.

16.2 RAISING OF CAPITAL IN THE PRIVATE SECTOR

This can be achieved by the use of a firm's retained profits, by bank loans, or by floating shares. If a company makes a profit it can retain this for development and expansion, or it can pay it out to its shareholders as a dividend, or it can retain some and pay out the rest as dividends. If it has cash flow problems of a short-term nature — that is to say that for a few months or even a year or two it needs more cash than it is likely to earn — it will borrow money from a commercial bank, repaying it with interest. Such techniques are usually employed for minor development or expansion by the firm.

However, if the firm wishes to start from scratch, or to expand in a large way compared to its existing size, it will offer shares for sale. Such shares can be *preference shares* (on which a given interest rate is guaranteed) or ordinary shares, on which all, or part, of annual profits are issued as dividends. Shares are also called *stocks* and *bonds*.

Suppose that a firm needs $1 000 000 to start, and issues 500 000 one dollar preference shares, with a 'guaranteed' 10% interest, and 500 000 ordinary shares of $1. If, after 1 year, profits are, say $200 000 it will pay $50 000 interest to preference shareholders and may pay anything from zero to $150 000 to the ordinary shareholders, retaining any balance for reserves. Of course, if it makes no profit, even the preference shareholders may receive no dividend. Assuming that it made $200 000 profit, and paid $100 000 in dividend to ordinary shareholders and retained $50 000, then in this case the ordinary shareholders would be receiving an interest of 20%, though the earnings per ordinary share are 30%. If the normal market rate for capital were 10%, say, the ordinary shares would increase in value, the amount depending on the future prospects of the company.

In all cases the firm will raise money by the sale of shares which, in our case, may be necessary to finance an engineering project, but the project must be shown to be economically viable in order for the shares to be sold.

16.3 RAISING OF CAPITAL IN THE PUBLIC SECTOR

Governments (which in this sense can include municipalities, commissions, or nationalized industries) are not normally in the position of floating shares, or of borrowing from commercial banks.

As has been stated, in developed countries major projects are usually financed from revenue derived from taxation or rates, although sometimes they are financed from loans. In developing countries, however, revenue funds are usually too small to cover any but very minor projects, and finance is raised from loans. Such loans may be restricted to the central bank (which can, if need be, recall money from commercial banks) or it may be raised by a public issue of government or municipal bonds. Such 'public' bonds are usually financed by public institutions, such as insurance companies or banks. These bonds have a guaranteed fixed interest rate, and are redeemed at agreed times.

A third form of such loans is that of contractor finance; in this case the contractor loans the money, which he may himself borrow, and government, in accepting his tender, agrees to repay it at an agreed interest over an agreed period of time.

Revenue funding is internal and need never be repaid, but loans, including contractor finance, have to be repaid. In the first instance, repayment can be achieved by floating more loans, but in the long run, all loans must be serviced from revenue, which will presumably increase, directly or indirectly, by the project earnings or savings. Even when construction is financed directly from revenue, funds used could have been directed elsewhere, so it is normal to consider the repayment of interest, and capital, on public as well as private sector porjects; in other words all engineering projects must be economically viable.

If government builds a toll road, income from tolls will service the loan. If, as is usual, the road becomes a public road, then the return of capital may be indirect. Either petrol tax may be increased, or, if not, national productivity will increase followed by an increase in national income, and in turn, by increased taxation revenue. In such cases, if the increased revenue exceeds the investment, then the economic viability is sound.

16.4 BASIC CONCEPTS OF ECONOMIC APPRAISALS

Costs involve those of design, construction, operation, and maintenance. Income comprises the revenue received from the project, and, in the case of government, the value of the social benefits, direct and indirect. These latter costs will be discussed later. Costs to government, but not to the private sector, may need shadow price adjustment. If the products of the projects are sold overseas for foreign currency, income may also need shadow price adjustment, but again not with the private sector which is paid in local currency. Thus it is apparent that an export project in the private sector could conceivably be sub-economic, but economic if it were a governmental project. In this case there would be a warrant for the government to subsidize losses, or to increase a firm's income by a grant-in-aid.

There is seldom one perfect solution to a design problem. Stage construction is a classic example: we can design a structure, be it a road or a whole town, and build it in one operation; alternatively we can build it piece by piece, enlarging or strengthening it as the need arises. The problem is to compare the economies of various possibilities which involve different amounts of expenditure over different periods of time. Further complications arise when we have to consider maintenance and income (or earnings), again of differing amounts over differing times.

This involves the concept of present day values, by which all incomes and expenditures are brought to a common day. At this day they are said to have present day values (PDV) or present values (PV).

If the market interest rate is $r\%$, then the value of D dollars today is $D[(100 + r)/100]^N$ in N years time; similarly the present value of D dollars, required or received in N years time, is $D[100/(100 + r)]^N$.

Using such formulae we can reduce all present and future incomes and expenditures to PDV on a common day.

16.5 THE EFFECT OF INTEREST AND LIFE ON COSTS

In all cases of costs it is assumed that the money is borrowed at a given rate of interest. If the project is everlasting, then the capital need never be repaid — the only capital costs will be the interest on capital used or borrowed. Conversely, if the project is assumed to have a life of say 20 years, the annual repayments (the fixed costs) will include both interest and capital repayment — that is to say they will constitute a mortgage.

Most civil engineering projects are assumed to have either long or infinite lives. For example, it may be estimated that a bridge will be required for 50 years, or a dam will take 50 years to silt up. If the cost is $1 000 000, using Table C for an interest rate of 5%, say, we can see that the annual repayments will be $1 000 000/18.2559 = $54 777, which is little more — only $4777 p.a. — than the interest repayments of $50 000, the annual cost applicable had the project been everlasting.

Had the interest rate been 10%, the annual repayments would have been $1 000 000/9.9148 = $100 859 as against $100 000 — less than 1% more. We can conclude by saying that the higher the interest rate, the less is the significance of the life of a project on its annual costs. Conversely it follows that the shorter the life of a project, the smaller is the importance of interest rates.

In the event it is apparent that in real life many projects have both limited and infinite life components. For example, if a factory is to be built, the building may be obsolete in 20 or 30 years time, but the land will still be there. Since there is little difference in annual costs over such fairly long periods, whilst the building and land costs could be separately assessed and summed, it would be normal to consider the life span of the major cost element and to use the same formula for both. That is to say, if the building cost were to exceed land cost, the total would be amortized over the 20 year building life. This is sufficiently accurate for most purposes.

16.6 SENSITIVITY OF LIFE FORECASTING

Since most engineering projects have lives in excess of 20 years, the effect of errors in the estimation of life expectancy is relatively small.

If a building costs $1 000 000, the interest rate is 10% p.a., and the forecast life is 20 years, annual costs (that is repayments of the loan), can be calculated using Table C as $1 000 000/8.5136 = $117 459. If the building were to last only 15 years the annual costs would be $131 473, and if it were to last 25 years, the figure would be $110 169; neither of these figures is catastrophically different from the estimate.

16.7 INTEREST RATES AND SENSITIVITY

The interest rate to be selected is the market interest rate, or the opportunity cost of capital. However, this is not always easy to determine.

In the private sector, the commercial bank overdraft rate is a guide — but this will be slightly less if the security of the loan is high, or if the entrepreneur has a good reputation. If possible, the entrepreneur will float a company and issue ordinary shares (called stock in some countries), forecasting a return interest on capital invested. This must be higher than the interest rate that would-be shareholders can get from investing in the commercial banks, building societies, and similar institutions with so called 'gilt-edged' stock, in order to make the purchase of such shares attractive.

In the case of government, loans can be floated at lower interest rates as the security is very high, and general revenue can guarantee payment, even if the scheme were to fail. If government floats a successful loan at say 7%, its next may be at 6%; if this fails to attract lenders it will increase the next to say 6.5%, and thus a fairly stable interest rate can be deduced and used in project appraisal. Conversely, in the private sector, several analyses should be undertaken at rates of the order of the commercial bank's lending rates, and the project sensitivity to different rates analysed.

When government floats a loan, the interest rate is fixed for the life of the loan. All bank loans and all mortgage agreements, however, enshrine reserved rights of the lender to change the interest rate over the course of time; thus a project which, if constructed, shows a profit when the borrowing rate is say 8%, may show a loss when this increases to 10%, and could go bankrupt. If such a company had floated shares it would give dividends (i.e. an interest to shareholders) unaltered by changing interest rates; but shareholders might prefer to sell their shares and invest elsewhere, if external interest rates increased, and share values dropped. However, the firm, like the government, will continue to function in these circumstances. Nevertheless there will be fluctuations in cash flow — a minor emergency may require short-term borrowing which may well be at higher rates. This again stresses the need, with all private sector projects, for analysis at different rates to check the sensitivities of the projects to changes in rates before commencement.

16.8   REPAYMENT PERIODS

Most engineering projects take time to design, construct, and commission, and this involves concepts of cash flow. The simple case is that of a person who wishes to build a house. Having obtained a building society agreement to finance his project which might take, say, ten months to build at $\frac{3}{4}$% interest per month, he will let a contract and borrow, at the end of each month, in order to pay the builder. Suppose his certificate is $2000 each month. At the *end* of each month he will borrow $2000. At the end of the tenth month he will owe the building society

$$\$2000(1.0075^9 + 1.0075^8 + \cdots + 1.0075 + 1) = \$20\ 689$$

He will then occupy the house, and start repaying *a month later*, when he will owe $20\ 689 \times 1.0075 = \$20\ 844$. *On this day he starts to repay the loan* over an agreed period, with equal monthly payments, until the loan is repaid. The period of delay occasioned by construction, and the delay between commissioning and income accrual, affects the magnitude of the debt.

In engineering project analysis we normally consider repayments on a yearly rather than monthly basis. This is more realistic as it simplifies calculations, also income from such projects may fluctuate from month to month, giving seasonal effects. Moreover it must not be forgotten that tax is payable on a yearly basis, although it is not paid on state or municipal projects.

For example, if a dam takes two years to build and costs $1 500 000 we might borrow $500 000 at the start and $500 000 every twelve months. If the interest rate were say 10% p.a., the day after completion we would owe

$$\$500\ 000(1.1^2 + 1.1 + 1) = \$1\ 655\ 000$$

and, when repayments were due *a year later*, we would owe

$$\$1\ 655\ 000 \times 1.1 = \$1\ 820\ 500$$

If the dam were assumed to be everlasting, the annual construction cost would be just the 10% interest due on the sum of $1 655 000 which was the accumulated debt the year previously, that is 10% of $1 655 000, i.e. approximately $166 000; whilst if we assumed a 50 year life we can calculate, using Table C, that the annual repayments would be $1 655 000/9.9148, i.e. $167 000.

As can be seen the difference in the two costs is negligible, especially if we bear in mind the uncertainties of the future half century! Indeed, it is because of such uncertainties that economic analysis can never be an exact science, and accuracies greater than 1% must always be suspect.

16.9   EFFECT OF INFLATION

No exact or standard technique makes allowance for future inflation, nor indeed would such a technique be valid as inflation rates cannot be predicted with any degree of certainty.

Basically we know that if productivity remains unchanged, inflation should have no effect on viability, since costs, measured in man-hours, will remain unaltered.

However, the increasing population may make land relatively scarce, and land values may increase *relative to productivity*. Again, materials such as fuel may escalate at a faster rate than productivity, and thus may increase maintenance or running costs, reducing profits and viability. If the construction has been financed on a commercial mortgage, with interest rates increasing with inflation, repayment rates will rise, reducing viability. Even future government bond interest rates may rise, and though fixed on a given project, they will affect its viability relative to similar new works. We can perhaps summarize by suggesting that inflation should normally be ignored, whilst mentioning the need to consider special cases.

Inflation will, of course, raise operational and maintenance costs; but it will also raise income from charges for future services provided. Again it must be assumed that project charges and income will increase with personal income, i.e. they will remain constant with productivity, and may thus be neglected in our calculations.

Obsolescence is perhaps similar in analysis in that it cannot be predicted with accuracy. However, a road network for a new city may well be more viable today than an electric mass-transit system; yet, in 20 years time the shortage and cost of fossil fuel may reverse the viability. As another example, a Channel tunnel for electric trains may be less viable than a road bridge in the seventies, but 20 years later the position could reverse when petrol is expensive and restricts road travel.

16.10   EFFECT OF TAXATION

In the public sector no taxation is normally due, as this would be robbing Peter to pay Paul; on the other hand, private sector investments are (or may be) complicated by taxation. However, the effect of taxation will depend on the taxation laws in the countries concerned. In many, if not most, countries companies will be required to pay company tax on profits earned, before dividends are distributed, but shareholders will not be required to pay further taxation on such dividends.

Again, whilst the amount of taxation payable and the time of its payment will affect income, the method of accounting will control the amount of taxable income. For example, if wear and tear were charged without allowance for the inflated costs of replacement, profits and hence taxation would be excessive in the early life of a scheme, and the developers could become bankrupt when replacement had to be purchased.

However, except in special cases, taxation is normally ignored in project appraisal since, whether private capital is invested in a factory project or in a bank, taxation will be equally payable on earnings.

PROBLEMS

Problem P16.1

At a period when gilt-edged government stock gives a 5% interest, a public company is floated with $1 shares. For several years a dividend of 10 cents p.a. is declared, and the share price at the Stock Exchange becomes constant at $1.20. Government then floats new stock at 7.5% interest rates and continues to do so. What is the likely effect on the company's share price?

Problem P16.2

An irrigation scheme which is long lasting involves the expenditure on completion of $1 000 000, when the market interest rate is 9%. Annual operating costs are forecast at $80 000, and sales of products at $250 000. Calculate whether or not the scheme is sensitive to interest rate changes.

Problem P16.3

A scheme with a 20 year life is completed at a cost of $1 000 000 with a market interest rate of 9%. Annual operating costs are $80 000 p.a., and income is anticipated at $200 000 p.a. On completion inflation commences at 10% p.a. affecting both operating expenses and income.

Develop a formula for net income in the Nth year, and then express this in terms of the present day value, taking the present day as completion day, and assuming year-end payments and accounting.

Chapter 17

Accounting Techniques

17.1 METHODOLOGY OF COST ASSESSMENT

Three basic techniques are available for the engineer in preparing a project appraisal. These are the payback period, the rate of return on book value, and the discounted cash flow techniques.

The Payback Period Technique

This technique involves the determination of the length of time between commencement of a project and the time when all investment has been repaid in benefit. *It is not recommended* as it ignores social benefits, the economic life of a project, and cash flow after the payback period is complete.

The Rate of Return on Book Value Technique

This is based on expressing accounting profit as a percentage return on capital invested, as measured by book values. This technique also ignores social costs, and assumes 'straight line' depreciation of assets (e.g. 10% value loss per year). It ignores present worth of assets which may disagree with book value, particularly in inflationary periods, and the fact that one item may affect the output of a whole series of others. *This method is also not recommended.*

The Discounted Cash Flow Technique

This method is the only recommended method for use. In this method, future expenditure and future receipts are discounted to 'present day' values. The underlying concept is that the investor is indifferent as to whether he has $100 now, or $110 in a year's time, if the ruling interest rate is say 10%, as the $100 invested now would fetch $110 a year later.

It is timely, however, to give a word of warning on the meaning of 'present day' as most engineering projects take many months, if not years, between commencement and completion. In such cases the cash flow may be more or less constant throughout the construction period, but income will only start some time after completion. Since, in engineering works we usually use annual calculations, the first income will become available one year after commissioning.

Any chosen 'present day' will allow the same answers to problems but careful selection will reduce the calculations involved. Suppose, for example, a toll road will take 5 years to build and is estimated to last 20 years; if the starting day is taken as the present day, then future income must be discounted to the completion day, i.e. from year 5 to year 0. If, however, the completion day is selected as the 'present day', 5 years of construction costs must be discounted *forward* to year 0 and income *backward* from the end of year 20 to year 0, a far simpler calculation. Such analysis is comparable with house building: the owner does not start to repay his mortgage until after the house is finished; rather he merely increases his debt by accumulating interest on his progress payments to the builder, and only starts repayment (in this case a month) after the house is finished.

It should be noted that in project appraisal it is tacitly assumed that the entrepreneur does not have idle capital; rather that he borrows money and repays it out of earnings of a project. Indeed, if in the unlikely event he had all the capital available before the start of a projects, he would invest this elsewhere, and withdraw it only as and when needed for progresss payments on his project.

Example 17.1

A project will take 1 year to complete at a cost of $100 000 payable at the commencement of the scheme. Returns will be $20 000 p.a., after each year's operation, and after 5 years use the scrap value will be $20 000. There is no inflation, and the interest rate is 10%. We have to determine the viability of the project.

It can be seen that the cash expenditure will be $100 000, and income will be $20 000 for 5 years, plus the scrap value of $20 000, that is $120 000.

Thus an investment of $100 000 will give a profit of $20 000 or 20%. However, if the entrepreneur had had the capital and had invested it in the market, he would have earned 10% of $100 000 after 1 year, and would be worth $100 000 × 1.7716 (using Table A) after 6 years, i.e. his profit from interest would have been $77 160, instead of $20 000. Thus it is apparent that the scheme is not viable.

To analyse the problem we must consider the concept of cash flow, based on PDV analysis.

Although it is not the simplest approach, consider the beginning of year 0, when the money is borrowed, as the present day. (Note that if the money is available and not borrowed then, the 'present day' would be the day it was withdrawn from investment at 10% p.a. interest, which amounts to the same thing.) We must now consider income. The first receipt is due 2 years after the present day, and

$$PDV = \$\frac{20\ 000}{(1+r)^2} = \$\frac{20\ 000}{1.1^2} = \$16\ 530$$

In other words a future income of $20 000 in 2 years time is only worth $16 530 at the present day. This is the same as saying that $16 530 today, invested at 10% p.a. will be worth $20 000 in 2 years time. By similar reasoning the PDV of the second receipt is $20 000/1.1^3 = $15 030, and the fifth receipt is $20 000/1.1^6 = $11 290 whilst the PDV of the scrap will be the same.

Table 17.1 PDV of cash flow

	Expenditure ($)	Income ($)
PDV of expenditure	100 000	—
PDV of first annual receipt	—	16 530
PDV of second annual receipt	—	15 030
PDV of third annual receipt	—	13 660
PDV of fourth annual receipt	—	12 420
PDV of fifth annual receipt	—	11 290
PDV of scrap value	—	11 290
	100 000	80 220

We can now tabulate the cash flow *in terms of PDV* (see Table 17.1). From this it can readily be seen that the PDV of income is less than that of expenditure, and the scheme is not viable. Indeed, the project would have to have cost less than $80 220 to make it viable and show a profit. The deficiency of $19 780 is 19.8% of the capital employed.

It should be noted that Table B can be used to ascertain the PDV of any individual future payment.

17.2 SELECTION OF THE PRESENT DAY

In Example 17.1, we selected the commencement of the project as the present day, but calculations can be simplified by a better choice of the completion day, as this enables us to use conventional mortgage tables without complication.

Using the completion date as the present day, the investor will owe more money on the borrowed capital, but the PDV of future income will be higher, as the future payments are now closer to the new 'present day'.

In the case of Example 17.1 the expenditure will now become $110 000, not $100 000, but the PDV of the first receipt will become $20 000/1.1 = $18 180 instead of $16 530.

Using Table C it can be calculated that the PDV of all annual incomes will be $20 000 × 3.7908 = $75 820. However, we still have to use Table B to determine the once only income from scrap as $20 000 × 0.62092 = $12 420.

Summarizing we can see that the PDV of expenditure is now $110 000 as against that of income of $75 820 + $12 420 = $88 240, a deficiency of $21 760, i.e. again 19.8% of the capital (now $110 000) employed.

Both this and the calculations in Example 17.1, are equally valid, but it is apparent that we need a new tool to accommodate both types of calculations — this is cost–benefit analysis.

17.3 COST–BENEFIT ANALYSIS

This is undoubtedly the most useful project justification tool available to the engineer. The benefit–cost (B/C) ratio is merely the *total* discounted estimated benefits divided by the *total* discounted estimated costs. In short-term projects *total* costs and *total* income are used, whilst in long-term (everlasting to all practical purposes) projects it is often more convenient to use annual amortized costs and annual income; the resultant answer is the same.

In the private sector, it is usually sufficient to consider merely financial costs, whilst in the public sector externalities must often be quantified in financial terms, and it is in this field that complications arise since it is difficult, if not impossible, to agree on a financial value of social benefits. This can be illustrated with the case of a road improvement and shortening. The costs can be forecast and are indisputable; the savings in costs to vehicles can be measured and quantified with some degree of accuracy; the amount of traffic growth which will be attracted by the new road is somewhat problematical; the value of the saving in driver's time is arguable. This is known for professional drivers, but for others it varies with income group and the type of traffic (at work or otherwise) as drivers cost their social time at a lower rate than their working time.

In Example 17.1, although income exceeded the costs of $100 000 by $20 000, we calculated that in terms of PDV, the income was only $80 220, from which it was deduced that the scheme was not viable. The B/C ratio of this scheme was simply $80 220/100 000 = 0.80, which, being less than unity, indicates that the scheme was not viable. Note that with the accuracy of economic forecasting, only two places of decimals, at the most, can be justified. The B/C ratio of the same scheme, recalculated in Section 17.2 with a different 'present day', was also 0.80, as it must be.

Suppose now that the anticipated scrap value had been $80 000, not $20 000. Then the *additional* income, at PDV timing of Example 17.1, would be $60 000/1.1^6 (or, using Table B, $60 000 × 0.56447) = $33 870. In this case total income, at PDV, would be $80 220 + $33 870 = $114 090, from which we can see that the B/C ratio would be 1.14, and that the scheme would be viable.

However, the figure of 1.14 is not very high, that is to say it is marginal; so we should examine the scheme's sensitivity in the light of its market uncertainty.

Its viability would change

(a) if the scrap value were to change;
(b) if the discount (market interest) rate were to change;
(c) if the future returns (income) were to change, even marginally;
(d) if taxation were involved;
(e) if an allowance were made for inflation.

Since many engineering schemes would show better B/C ratios, such a scheme in private enterprise, where externalities of social costs and benefits (if any) are ignored, would receive low priority.

Example 17.2 Development of a mine

In an underdeveloped country a project is proposed to mine ore. Due to political risks, the borrowing rate is high at 10%. The estimated costs are the construction of the infrastructure ($10 million), payable at the start of the job, and $10 million for machinery payable after 1 year. Production is planned to commence after 2 years, and to last for 5 years, after which the scrap value of machinery is negligible. Operational labour costs are forecast at $2 million p.a., and sales at $10 million p.a. We have to determine the cash flow, assuming repayment of loans as soon as possible, and the B/C ratio, assuming year-end payments and income.

Table 17.2 shows the cash flow. Each year the interest due during the previous year has been added and, in the 5 years of production, surplus income has been used to repay the outstanding loan. A profit of $3.4 million has been shown for the first time after the sixth year; this could then be paid out in dividends, with a final dividend of $8 million when the scheme is complete. Such dividends are due to the entrepreneur, the mortgagee having been repaid.

For the expenditure of $38.6 million, including 10% interest on balance, an income of $50 million has been achieved, but $50/38.6 = 1.30$ is *not* the B/C ratio, as income and expenditure have not been brought to present day values.

We can now consider Table 17.3 of present day values, taking, for convenience, the present day as the end of year 2 when construction ceases and income accrual begins. At this time the money borrowed has acquired interest and is $10 million $(1.21 + 1.1) = \$23.1$ million. The discounted annual income and annual operational cost are obtained from Table B.

Table 17.2   Cash flow

End of year	$millions Expenditure	Receipts	Balance	Remarks
0	10	—	− 10	Work starts; infrastructural costs
1	10 + 1	—	− 21	Interest on infrastructure plus machinery costs
2	0 + 2.1	—	− 23.1	Work finishes, income starts; total expenditure plus interest
3	2 + 2.3	10	− 17.4	Labour expenditure, interest expenditure; income
4	2 + 1.7	10	− 11.1	—do—
5	2 + 1.1	10	− 4.2	—do—
6	2 + 0.4	10	+ 3.4	—do—
7	2	10	+ 11.4	Close down
Total	38.6	50	+ 11.4	(Profit)

Table 17.3   Calculation of PDV of a mining project ($millions)

End of year	Loan	Discounted net operational costs	Discounted net income	Remarks
2	23.1	—	—	'Present day'
3	—	1.82	9.09	PDV of future sum
4	—	1.65	8.26	PDV of future sum
5	—	1.50	7.51	PDV of future sum
6	—	1.37	6.83	PDV of future sum
7	—	1.24	6.21	PDV of future sum
Total	23.1	7.58	37.90	$B/C = \dfrac{37.90}{23.10 + 7.58} = 1.24$

In considering Table 17.3 it can be seen that at completion of the project it will have cost $23.1 million, plus a commitment of $7.58 million at present day values for operation, and an expectation of sales (benefits) of $37.90 million, also at present day values, giving a B/C ratio of 1.24.

The problem could have been solved more simply using Table C, which gives a factor of 3.7908 for 5 years at 10% interest, from which we could deduce the PDV of future operational costs to be $7.58 and of income to be $37.90 millions.

It should be remembered that the B/C ratio has been defined as the ratio of *all* discounted benefits to *all* discounted costs. There is a tendency, incorrect by this definition, to consider the *nett* income, 10 less 2, i.e. $8 million in this case, and to discount this to PDV. This would give a figure of $30.32 million, and a B/C ratio of 1.31. It will still be positive, but will not comply with the above definition of the B/C ratio.

17.4   EFFECT OF PROJECT LIFE

In Example 17.2 we considered a project with a 5 year estimated life. Had it been possible to extend the life without any increase in capital investment, still at a market rate of 10%, the PDV of both operational expenses and income would have increased. Using Table C we can obtain factors of 3.79 for 5 years, 6.14 for 10 years, and 10.0 for an infinite life. From such data Table 17.4 can be constructed, with the investment at PDV of $23.1 million.

It can be seen that when life exceeds some 20 years the increase in the B/C ratio is marginal. This applies to many engineering projects, and certainly to almost all civil engineering ones. It is particularly true if one accepts that in the real world maintenance costs are likely to increase with age, which would further reduce the B/C ratio at long life.

In the case of the infinitely long life project, if we took the annual cost, being the annual interest on capital of $2.31 million plus the annual operational cost of $2

Table 17.4 Effect of life on viability of a mining project ($million)

Life of project (years)	PDV of income	PDV of operational costs	Total costs	B/C ratio
5	37.90	7.58	30.68	1.24
10	61.40	12.28	35.38	1.74
15	76.06	15.21	38.31	1.99
20	85.14	17.03	40.13	2.12
25	90.77	18.15	41.25	2.20
30	94.27	18.85	41.95	2.25
00	100.00	20.00	43.10	2.32

million, and compared it to the annual income of $10 million, the B/C ratio would still be 2.32. Thus, on long lasting projects we can calculate the B/C ratio either by considering total discounted construction costs and discounted net receipts, or we can consider total annual incomes and total annual costs.

With long life projects, it is convenient and simpler to calculate B/C ratios from annual cost data, rather than from data discounted back to present day values.

PROBLEMS

Problem P17.1

A developer, with a plot of negligible value, calculates that he can build a house and rent it for $4000 p.a., payable in advance, and can borrow money from a building society at a mortgage rate of 10% p.a., payable in arrears. He can build the house himself in 3 years, at $10 000 payable at the beginning of each year, or by contract in 1 year for $35 000, payable on completion. Which scheme is the more profitable?

Problem P17.2

In the above schemes all rented income is to be devoted to paying off the bond. Approximately how many years would it take in each case?

Problem P17.3

It is calculated that a machine costing $10 000 will last 5 years giving an output valued at $2000 p.a. After 5 years it has a replacement value of $5000. Caclulate the B/C ratios at market investment rates of 5% and 10% respectively, assuming year-end payments.

Chapter 18

Decision Making on Cost–Benefit Analysis

18.1 THE SENSITIVITY OF COST–BENEFIT DATA

It is obvious that a scheme is viable if the B/C ratio exceeds unity, but the entrepreneur is likely to be unwilling to embark upon a scheme which complies with this rule unless he is satisfied that, if future predicted conditions change, the project will still remain viable. In other words he needs to know how sensitive the B/C ratio would be to possible changes in input parameters. Whilst the most important of these is likely to be the interest rate, others will be the construction costs, the market price for the product, the maintenance forecasts (which may include the labour rate), the forecast obsolescence period, taxation, and inflation. Only when he has the gamut of such information is he likely to proceed.

18.2 THE MARKET INTEREST RATE

Engineering projects are likely to last several years, during which period the market interest rate is likely to change — particularly since, with time lags, it is associated with the bank rate. If the entrepreneur borrows from a bank or other commercial institution, he is unlikely to be able to do so at a fixed rate; for example, all mortgages and loan agreements give the lender the right to change the rate during the life of the loan. Even if he floats shares for a company, the investors are concerned with prospects *vis-à-vis* alternative investments, so the same principle applies.

Again let us consider the viability of the 5 year mining project in Example 17.2, which had a B/C ratio of 1.24 when the market interest rate was 10%, if this changed to 5% and to 15%.

At 5% the capital debt at the present day reduces to $21.5 million, the PDV of future operational costs increases to $2 \times 4.3295 = \$8.66$ million, whilst that of sales increases to $10 \times 4.3295 = \$43.30$ million, giving a B/C ratio of

$$\frac{43.30}{21.5 + 8.66} = 1.44$$

At 15% interest the investment increases to $24.7 million, and using a factor of 3.3522 from Table C, PDV of operational costs and income reduce to $6.70 million and $33.52 million respectively, resulting in a B/C ratio of

$$\frac{33.52}{24.7 + 6.70} = 1.07$$

In the above (and other) calculations it has been presupposed that the market interest rate and the loan rate are the same thing, and are changed simultaneously. Of course, it could be that money was borrowed for the scheme at a per cent or two above the market rate, which would necessitate using different rates for the loan and the income, which would slightly affect the B/C ratios.

It can be seen that since interest rates could well lie between these limits, according to political stability, the scheme is still viable. It can readily be calculated that short-term schemes are less affected than long-term ones by changes in interest rates, and that in long-term schemes, an increase in interest rates may negate a scheme's viability.

18.3   EFFECT OF CHANGE OF FUTURE INCOME

Schemes which result in products, such as luxury goods for which the demand and price may vary considerably with the business cycle, need higher B/C ratios to make them attractive. Such 'products' include mining projects, since mineral prices tend to be extremely cyclic.

Again, in Example 17.2 income was $10 million p.a. less $2 million operating costs. It can be calculated that only a 20% drop in the sales price of the ore would result in a negative B/C ratio. Such a scheme would only be a viable risk if a long-term rise in ore price were anticipated. Higher B/C ratios are always necessary for mining projects due to the cyclic nature of ore prices, and the scheme having such a low B/C ratio would only be viable if the income were based on anticipated prices for ore at the trough of a business cycle.

However, the scheme was not so sensitive to labour costs. It can be calculated that even a 50% rise in these, to $3 million p.a., would only cause the B/C ratio to drop to 1.10, and a 50% increase is likely to be extreme. Of course, it could occur if inflation were high, but in such circumstances it could be anticipated that the ore price would also rise.

18.4   EFFECT OF ERRORS IN CONSTRUCTION
COST ESTIMATES

Consultants can usually forecast construction costs within 10%. Thus the B/C ratio is unlikely to be markedly affected by poor forecasting of such costs. The same limits of accuracy should also apply to operational and maintenance costs, but again it is more difficult to be accurate with mining than with other industries.

18.5 EFFECT OF TAXATION

Firstly, it should be borne in mind that profit from all private enterprise schemes is liable to taxation, so to some extent, we have a self-balancing situation. Even if the entrepreneur had had the capital available and had invested it in the market instead of on the project, he would have had to pay tax on the interest earned.

A lot depends on the presentation of accounts (the statement of affairs, the balance sheet), particularly as to which expenses are deducted annually to offset income. If we consider Table 17.2 for the scheme in Example 17.2 taxable profits begin only in the sixth year, the majority being in the seventh; and if tax were payable only then, its value discounted to the PDV would be relatively small. However, income starts in the third year and will normally be taxable, as it will in all subsequent years. If the tax assessor will then accept a 20% p.a. write off for depreciation (i.e. $4 million), the taxable amount will be $10 million income, less $2 million (operational expenses), less interest payable (which depends on whether share or other capital had actually been available to the entrepreneur, or whether it had been commercially borrowed), and less $4 million for depreciation.

The problem is magnified, since in reality the company tax rate cannot be forecast for future years, and it is thus normal to ignore the effect of taxation on B/C ratio calculations, again stressing the approximate nature of such ratios.

18.6 EFFECT OF DEPRECIATION AND OBSOLESCENCE

All engineering plant and public works depreciate with use, however well they are maintained.

Firstly there is wear and tear: a gradual, usually accelerating, process of physical deterioration. Sooner or later the time will come when it will be more economic to scrap the item and replace it, rather than to repair it.

Secondly there is obsolescence. Diesel locomotives may last 50 years before they become uneconomic to maintain and need replacement; but if there is an escalation in the price of diesel fuel, it may be economic to scrap them after only a few years and to replace them with electric locomotives. Obsolescence is certainly a feature of transport systems, because transport technology advances quickly. Airlines with propeller driven aircraft were compelled to sell them cheaply, even though physical deterioration was minimal, when rival airlines re-equipped with jets. Similarly, excellent ocean liners, such as the *United States*, were scrapped after some 20% of their physical lives, as they could not compete with the subsonic jets introduced a year or two after their introduction.

Yet, in spite of these difficulties, we must know how much to depreciate our assets each year, not only to prepare our statement of affairs (balance sheet) for taxation and other legal purposes, but also to ascertain how much depreciation to allocate to the production costs of our output. Such depreciation money must normally be set aside for replacing the machinery, or works, in due course, as a *sinking fund*. Thus, with his greater technological knowledge, it behoves the engineer to assess the life of his machinery or works. It is seldom that he will be able to forecast obsolescence, and he will usually work on estimates of physical deterioration, even though these are vague, as life can often be extended with increased maintenance.

Luckily, most engineering works and even machinery, are long lasting. It makes a great deal of difference in economic terms if a car lasts 2 years or 4 years (does a car depreciate at 50% or 25% p.a.?) but less if a lathe lasts 10 or 20 years (10% or 5%) and even less if a bridge lasts 50 or a 100 years (2% or 1%).

In traditional (historic) accounting a 'straight line' depreciation is assumed. If a lathe costs $10 000 and is given a 10 year life, the company's annual statement of affairs will show $1000 a year depreciation in the debit, or expenditure, column. This will be deducted from before-tax profits. Also this fixed cost will be added to production costs; if 1000 items a year are produced, $1 will be added to the fixed cost of each, producing an extra income of $1000 a year, assuming the demand still existed.

The question resolves itself into determining how depreciation and obsolescence can be handled in B/C analysis. Once a life has been assessed for a project, the answer is that depreciation can be ignored in B/C analysis, since in all calculaitons income will cease after this life, and the investment will have become valueless. In the case of obsolescence, however, since this cannot be estimated at the start of a project, no allowance can be made. An exception might be the case where goods, such as propeller driven aircraft, with a normal life of say 10 years, might have to be ordered to fill a gap before jet replacements, on the drawing board, become available; in this case of planned obsolescence, they will be written off to a low value in a short period in the B/C analysis.

Of course, any sum set aside for depreciation is offset against profits for taxation purposes; this can complicate an analysis, particularly as interest earned by a sinking fund is taxable.

Digression: Sinking Funds. If a firm borrows money for a lathe, say, it will normally assess the life of the lathe and pay back the money as a 'mortgage' over its life, and production 'on-costs' will be the mortgage repayments.

However, if the firm had used its own funds, it would have had to set aside a *sinking fund*, so that when the machinery beame worn out it could be replaced.

In practice the sum allocated to depreciation is not paid out as profits, but is invested. The investment can be within the firm (in which case no tax is paid *directly* on the interest), or outside (in which case tax is paid by the investor directly on the interest received). Suppose a lathe costs $10 000 and has an estimated life of 10 years. Since $1000 a year, invested at interest, would yield more than $10 000 after 10 years, a smaller sum than that will need to be set aside to accumulate the $10 000 for the replacement.

The fund set up for this purpose is called a sinking fund. Let x dollars be put aside each year and invested at $r\%$, where the project is written off in y years. Then after n years $\$x$ will have been invested for $y - n$ years, and when the project is written off $\$x(1 + r)^{(y - n)}$ will be available.

Using Table A for our example of a $10 000 lathe for a 10 year life at, say, 5% investment we can say:

Value of investment $x made after 1 year, after 10 years = $x × 1.5513
Value of investment $x made after 2 years, after 10 years = $x × 1.4775
Value of investment $x made after 3 years, after 10 years = $x × 1.4071

— etc. —

Value of investment $x made after 9 years, after 10 years = $x × 1.0500
Value of investment $x made after 10 years, after 10 years = $x × 1.000

Therefore

Total sum available after 10 years = $x × 12.5579

But $10 000 is required. Thus

$$x = \$795$$

Note that with year-end balance, the first sinking fund payment is made after 1 year, and invested for 9 not 10 years, whilst the last payment earns no interest. Note also that this value can be determined directly from Table D, where the factor is 0.07950. Thus if a sinking fund of $795 is made after every year and invested at 5% p.a., $10 000 will then be available for a replacement.

The problem is what depreciation should be charged against production and entered in the annual statement of affairs.

This should still be $1000, because (if straight line depreciation is valid) this will be the true loss each year. However, since only $795 is set aside as a sinking fund for investment, the profits of its interest earnings could be subject to taxation, i.e. taxation on the profit of an investment. However, this is an oversimplification, the details of which do not concern us, as the interest earned will increase each year from zero in the first year's operation (when no money has been involved) to a maximum in the tenth year, when 9 × $795 will have been invested and will be earning 5% p.a. interest.

Digression: Amortization. This is a more usual and convenient method of assessing the fixed costs as a charge against production, and is akin to a mortgage.

Amortization is defined as the extinguishing of a debt by means of a sinking fund. In the previous example of a sinking fund, by setting aside $795 after each year and investing it at 5%, after 10 years the initial sum of $10 000 became available to replace the worn-out plant. However, had the $10 000 orginally been invested in the market at 5%, we would not have had $10 000 but $10 000 × 1.6289 (from Table A) = $16 289 available. Had $10 000 been borrowed in the market at 5% interest rate, repayable in a lump sum after 10 years, a sinking fund of

$$\$\frac{795 \times 16\ 229}{10\ 000} = \$1295 \text{ p.a., not } \$795$$

would have been necessary to give $16 289 in 10 year's time; this fund would be due to the lender of the $10 000 capital. However, if the $10 000 capital been borrowed on a mortgage basis, with repayment of the loan in equal annual instalments, the investment would have been amortized. Here, the reciprocal of Table C is used. The amortization (annual payment to write off this loan) would be

$$\$\frac{10\ 000}{7.7217} = \$1295 \text{ p.a.}$$

Thus the annual depreciation write off of $1000 p.a. leaves a cost shortfall of $295, which represents interest on capital, and so $1295 is a necessary and legitimate charge to production costs, although $295 p.a. could be taxable.

18.7 COMPARISON OF DEPRECIATION METHODS

At this stage it is interesting to see the effect of life on amortization repayments. Again referring to the $10 000 lathe, and using Table C, it can be seen that our annual payments for amortization will be $10 000/7.7217 for a 10 year life at 5%, i.e. $1295 p.a., or $10 000/6.1446, i.e. $1627 p.a., at 10% interest. If the lathe now lasts 20 not 10 years (and the amortization is arranged over this period) $802 and $1175 are the annual charges respectively.

Note that doubling of life at 5% interest gives only a 38% reduction in amortization costs, and at 10% interest, only a 28% reduction. This illustrates that life estimates are not so critical as would first appear, particularly with long life and high interest rates.

18.8 SELECTION OF DEPRECIATION METHOD

In the example, with a $10 000 lathe lasting 10 years, we have the choice of recording depreciation as $1000 p.a., $795 (a sinking fund invested at 5% p.a.) or $1295 p.a. (amortized at 5% p.a.).

If $1000 p.a. were charged to production costs, it would appear at first sight that the output would be overpriced, it being more logical to charge $795, since patently any sinking fund contributions would not stand idle. However, in this case, it is assumed that the entrepreneur would not only have the $10 000 working capital (before the projection started and after it ended) *but that he would require no return on it*. Obviously the alternative to the project would have been investment, so by only charging $795 or even $1000 p.a. to production, the entrepreneur might have made a profit (dependent upon sales price) but he would be losing a profit on his capital to which he is entitled, and which shareholders would demand. It follows that the correct method is to charge the amortized payment of $1295 p.a. to production. By doing so, the entrepreneur can either borrow capital from a financial house and repay it with interest, or he can give interest to his shareholders if he raises the capital by a sale of shares.

Taxation, and the showing of depreciation on the annual statement of affairs, is another matter. Many tax and legal authorities may require only the $1000 p.a. to be shown; thus, since a further $295 is being spent, or put aside, for interest, the entrepreneur will be paying tax on this. However, in the long run tax has to be paid on all the firm's profits defined as the surplus of total income over total costs; whether this originates from the choice of a depreciation figure, or from income earned on capital, is immaterial.

It must, however, be borne in mind that many public works, e.g. roads, sewage, water supplies, harbours, airports, railways, and the like, are undertakings of the state, the municipality, or quasi-government commissions which pay no tax, and should operate on a no-profit, no-loss basis; so tax complications are often irrelevant to the engineer, and, in other cases, are primarily the concern of the accountant. In the event, in the real world we are faced with problems of inflation, where the historic (past) value of equipment, or depreciation is irrelevant to present values.

18.9   EFFECT OF INFLATION

In the example of the lathe we have considered the case of zero inflation. If we assume that inflation exists we cannot reliably assess a figure for it: at the time of writing national inflations vary between some 10% and 30%. Inflation will patently not be at the same rate for all items: if it were, fewer problems would arise. However, even if we do estimate inflation over the life of engineering projects with any accuracy (such projects often last 10 to 40 years in service) we must assume, for simplicity, that it will be equal for all goods and resources. It should also be remembered that labour (and in developing countries the training of unskilled labour will increase individual productivity, which could increase faster than inflation, though the degree is again difficult to assess) represents productivity, as well as price per hour, so costs of items, measured in labour units, should not increase; rather, if anything, they should decrease.

Assume, for a moment, that inflation rises at rate I. If future house rentals were, in inflationary periods, fixed in price (say by law — governments seek to obtain votes by fixing rents, a policy which redistributes wealth and mitigates against housing development in the private sector) then *future* rental payments in *real* terms will be less, as they will have less purchasing power than they have at present.

The PDV of any future payment, n years hence, is $P/(1 + r + I)^n$, but, if P, the income, is not fixed by law it is liable to increase at, or near, the inflation rate I, so the PDV of any future payment becomes $P(1 + I)^n/(1 + r + I)^n$, which is approximately the same as the original value $P/(1 + r)^n$. Thus, in this (the normal) case, inflation can be ignored in our analysis. However, this is an over-simplification in the taxable world when annual statements of affairs, which probably may not be adjusted for inflation, have to be produced. Again, there is always a time lag between the inflation of costs, and of income, a good example of which is railways. On a state railway fares might be stabilized at a point where the B/C ratio is unity, i.e. annual costs equals annual income. Then, with inflation, costs rise, but there is liable to be a delay of a year or two before government will allow fares to be raised, so a loss is shown, a subsidy is called for, and the management is then criticized!

Finally, it should be noted that the *real* rate of return on capital, where R is the rate after tax, is $[(1 + R)/(1 + I)] - 1$; but since, in present conditions, it is seldom possible to invest capital at a tax-paid rate exceeding inflation, the real rate is less than unity, i.e. we can make no profit by investing at 20% interest, on which we pay 50% tax, if the inflation rate exceeds 10% — by doing so we merely *reduce* the loss of our future purchasing power. If this is analysed it means that it may pay the entrepreneur to borrow money for development, but equally the finance house should not lend capital, as it would show a loss in real terms. Of course the finance houses themselves can only operate on money which they borrow, in turn, from smaller investors, and the only motivation at present from the smaller investor is the need to minimize his loss on capital! For this reason investment loans are often difficult to come by, and entrepreneurs (company directors) tend to withhold profits, and to 'float' extra share capital, to finance development. However, in all cases, the same principles of cost analysis should be applied, using the market interest rate to discount future costs and incomes to present day values.

18.10 ENTREPRENEURIAL PRIORITY DECISIONS ON BENEFIT–COST RATIO TECHNIQUES

The entrepreneur has to have B/C ratios calculated on several differing assumptions, the most important being the interest rate, the income (sale price may be controlled by laws of supply and demand) and, to a lesser extent, the project life. Any project giving B/C ratios exceeding unity in all, or nearly all, combinations of input is viable, but all involve risk elements, so projects with B/C ratios marginally in excess of unity will seldom be considered.

Table 18.1

Scheme	Capital required ($ millions)	Probable B/C ratio	Estimated life, start to finish (years)
A	1.5	1.6	5
B	3.0	2.1	10
C	2.0	1.8	15

The entrepreneur will not put all his eggs in one basket; he will investigate many schemes and choose the one, or more, which have the highest B/C ratios. Suppose he were considering three schemes as in Table 18.1. If $6.5 million were available or could be borrowed, he might commission all three schemes; if only $5 million were available, schemes B and C; if $3.5 million, scheme B only; and scheme A, least profitably, if only $1.5 million were available. Profits from this might enable scheme C to proceed, in turn subsequently allowing the more attractive scheme B to be commissioned. Thus availability of capital, as well as profitability, will influence choice and priority.

18.11 ALTERNATIVE TECHNIQUES FOR ASSESSING PROJECT PRIORITIES

Since benefit–cost analyses are often lengthy, the entrepreneur may wish to use simpler tools for selecting the priority of adoption of alternative schemes. Two such tools are the *internal rate of return*, and the *rate of return on book value* methods. These are best studied by reference to Example 17.2 where a mine, with a 5 year life requiring $10 million investment at the beginning of each of 2 years, is expected to yield $8 million nett after the end of the third and each of the subsequent 4 years.

The internal rate of return (IRR) method entails the establishment of the discount or interest rate which would result in the present day value of nett income equalling that of the investment.

Thus, in the example, taking the present day as the completion date after the end of 2 years, and R as the interest rate, we can obtain the following equation, in millions of dollars:

$$10(1 + R)^2 + 10(1 + R) = \frac{8}{1 + R} + \frac{8}{(1 + R)^2} + \frac{8}{(1 + R)^3} + \frac{8}{(1 + R)^4} + \frac{8}{(1 + R)^5}$$

Solution, by trial and error, gives a figure of between 17% and 17.5%, i.e. 17% for all practical purposes.

It should be noted that this rate of internal return is the interest rate at which the B/C ratio is unity. Thus, even if a project were to have scrap value, its PDV can be estimated, and an IRR value obtained.

The rate of return on book value is the interest rate obtained by expressing the accounting profit as a percentage return on the capital invested, this capital being a 'book value', assuming straight line depreciation.

In this example Table 18.2 would be compiled (in $ millions) as shown. From this it can be seen that the rate of return is

$$\frac{2.86}{11.43} = 25\%$$

Whereas the IRR value of 17% is real, in the sense that if the opportunity cost of capital is less than 17% the scheme is viable (i.e. the B/C ratio exceeds unity), the

Table 18.2

| Year | Nett revenue | Depreciation of scheme | Nett profit | Book value of scheme | | |
				At start	At end	Average
1	0	0	0	10	10	10
2	0	0	0	20	20	20
3	8	4	4	20	16	18
4	8	4	4	16	12	14
5	8	4	4	12	8	10
6	8	4	4	8	4	6
7	8	4	4	4	0	2
Total			20			80
Average			2.86			11.43

rate of return on book value is misleading, since this is not a true percentage indicating viability. Indeed the only use of this method is as a tool to assess priority — a scheme with 40% might take priority over one with 30%.

PROBLEMS

Problem P18.1

A project in the private sector will cost $10 million, and will last 15 years. Of the $10 million half will be spent on an infrastructure with an indefinite life. Annual operating costs are estimated at $1 million, and sales at $3 million. The market interest rate is 10%, but could well rise to 12%. Comment on the viability of the project if it were: (a) a mine; and (b) an agricultural development, ignoring any time taken for construction.

Problem P18.2

Discuss the change in viability of the scheme in Problem P18.1 if inflation occurred with the market interest rate rising to 12%.

Problem P18.3

An irrigation scheme and agricultural project is proposed which will have a long life and cost $10 million, including $5 million for an infrastructure, the market interest rate being 10% due to risk. Operating costs and sales are estimated at $1.2 million and $2.5 million respectively. Government can borrow money at only 6.0%, due to the security it can offer. Calculate the B/C ratios for private and for public enterprise, and comment on the viability, ignoring construction time.

Chapter 19

Cost–Benefit Analysis in the Public Sector

19.1 DIFFERENCES IN FINANCIAL APPROACH

In the private sector the motive for a project is the maximization of profit, whilst in the public sector the motive is, ultimately, the raising of living standards. The entrepreneur seeks to have an IRR well in excess of the market borrowing rate; he chooses a scheme with the highest B/C ratio. He will pay tax on profits. The public sector, on the other hand, pays no tax and can work on lower borrowing rates, since government (including municipalities) can borrow, externally or internally, and can offer security in that it can repay loans even if a scheme proves not to be a paying proposition. Thus a scheme which may not be profitable under private enterprise may be viable in the public sector.

Again a municipality may decide to use land for public purposes and not charge for the use of such land; car parks often fall into this catagory. Also government may be concerned with foreign exchange and may use shadow pricing in connection with benefit–cost analysis. Government might even proceed with a project which has a *financial* B/C ratio of less than unity in order to create employment, or to produce subsidized goods which will make other industries, which may be private, viable. Finally, government is concerned with other externalities such as social costs, which are seldom the concern of the entrepreneur. In these cases the *economic* B/C ratio should exceed unity.

19.2 SOCIAL COSTS

These can be economies or diseconomies. The opening of a mine in an underdeveloped country will involve the construction of an infrastructure — usually a road or a railway. Such a system will seldom be reserved for the exclusive use of the mine, and it will be used by others. Perhaps, before it was built, people walked along tracks — after a road is built they can travel more quickly and their productivity would increase. Local land values will increase, as crops can be brought to market. However, in a developed country the construction of an international airport or even of a restricted-access freeway, will lower land values in certain adjacent zones, and raise them in others. The engineer and economist must quantify such social costs (or gains) in financial terms, and use them as inputs to benefit–cost analysis. It is really the inclusion of such costs which differentiates between a simple project appraisal, and a benefit–cost analysis.

This poses the problem of how to quantify such 'costs', that is social losses or gains, in financial terms.

19.3   TECHNIQUES OF ASSESSING SOCIAL COSTS

The engineer, or economic analyst, has two techniques (tools) available to assess social costs: theoretical assessment, or the application of data from previous local or foreign schemes.

Theoretical assessments rely on his judgement in that, as the result of a scheme, he will estimate such aspects as the change in population, the change in its productivity (earning capacity), the change of land values, etc. If, for example, a paper mill is to be established in an undeveloped area, he will need climatic and soil reports to indicate the tree growth potential, from which he must not only estimate supply of timber, but also the changing land values. He must then estimate any similar effects for the changing position of nearby land – whether, for example, a dairy industry will be established to provide for the mill and forestry workers.

He will, of course, have to estimate plant and labour costs, the mill's output and the value, *in situ*, of the output. He will next have to estimate effects of the infrastructure on land values between the mill and market and even on land beyond the mill, the occupiers of which may now also use this road or rail for some of their journey to a market.

Last, but not least, he will have to estimate the effects of disposal of poisonous liquid by-products generated by the mill. If, for example, they are discharged into a stream, fishing downstream will stop, and the ecological balance will be destroyed. If such water is used, downstream, for agriculture, industry, or for drinking, the diseconomies must also be estimated in monetary terms.

The engineer will obviously seek to apply data from similar projects. The study of such projects, usually overseas, must be undertaken — the consulting engineer is well placed to obtain such data — and the social and financial costs broken down and recalculated for the different environment. In both theoretical and applied techniques much learned or experienced judgement will be required, but all such judgement will involve the time element, for schemes developed today will last for perhaps 10, or possibly for 100, years, but their effects may last for ever. Some people argue that we have no right to commit future generations, but these doubtful arguments are social, not engineering, and are certainly open to question. At present we are wasting the energy resources of our children — albeit with guilty consciences — for today's convenience. As has been seen the present day effects of long-term costs and incomes of schemes which last more than some 20 years are small and the long-term effects on economic viability can normally be ignored. It can be concluded that, whatever the background, feasibility studies in the form of cost appraisals, including cost–benefit analysis (CBA), are necessary, and the first requirement, having determined the applicable interest rates, is to decide on the accounting values to be employed.

There are, of course, cases where foreign data are not available, and the analyst will have to start from scratch and quantify data on the basis of logic. For example, if a suburban rail link is planned, he will have to estimate the demand for the service, the reduction of land values immediately adjacent to the track, and the increase in values resulting from time-saving, in property a little further off. The analysis, in such cases, should be repeated with various inputs to check its sensitivity.

In such a case the effect on the viability of existing transport modes must also be postulated. How much custom will be drawn from existing bus routes? How much will this affect service frequency and viability? Concomitantly, such a case, with a growing population, raises the case of the alternative to the rail project, if the existing road infrastructure cannot cope with future traffic growth.

19.4 SPECIFIC PROBLEMS OF QUANTIFYING SOCIAL COSTS

Social costs will vary with the earning capacity of the persons affected, with the size of the economy or diseconomy, and with the time at which they apply. A good illustration is that of a road improvement. Obviously the managing director will value time saved more than, say, his office cleaner will. However, both will value their time more if it is saved during working hours, than if saved during private time. Again, neither will assess the saving of a few seconds, but will value the saving of a few minutes, and may value the saving of an hour's travelling time, even more, *pro rata*, to the time saved. Thus the analyst cannot take costs per hour from one situation, and apply the data blindly to another. Yet it is such input data which are required as a major input as a benefit on transportation improvement schemes.

Example 19.1

A rural road improvement is proposed, shortening the existing 15 km route, from A to B, to 10 km. Road construction costs $100 000 per km, and maintenance $5000 per km per annum. The acquisition costs of land for the new route are $50 000 per km, twice the estimated sales value of the old route. The road authority can borrow money at 5% p.a. The average vehicle travels at 100 km/h at a cost of 10 cents/km, and has one occupant. No traffic growth is envisaged. What is the minimum traffic count to make the scheme viable if occupants value their time at: (a) $6 per hour; and (b) $3 per hour; assuming that construction time is ignored?

It is advisable to tackle this problem in three stages: initial costs, nett income, and B/C ratio.

(a) *Initial costs*

	$
Construction 10 km at $100 000	1 000 000
Land acquisition less sales of old land, $10 \times \$50\,000 - 15 \times \$25\,000$	125 000
Total initial costs	1 125 000

The new route will last indefinitely; thus the investment need not be repaid, only the interest. Therefore

Annual costs of initial investment, 5% of $1 125 000 $56 250

(b) *Nett annual 'income'*

Assume that N vehicles per annum use the road, 5 km saved at 10 cents/km; then nett savings on vehicles are

$$N \times \frac{5 \times 10}{100} = \$\frac{N}{2}$$

Time saved per vehicle per trip:

$$5 \text{ km at } 100 \text{ km/h} = \frac{1}{20} \text{ hours}$$

Therefore cost savings in time at $\$y$/h are

$$\$\frac{N \times y}{20}$$

Maintenance savings are

$$5 \text{ km at } \$5000 = \$25\,000$$

Therefore total savings per annum are

$$\$25\,000 + \$\frac{N}{2} + \$\frac{Ny}{20}$$

(c) *Benefit–cost ratio*

For minimum justification the B/C ratio must equal unity. Since the project has an infinite life, then

$$\text{B/C ratio} = \frac{\text{annual benefits}}{\text{annual cost}}$$

Therefore (in dollars),

$$25\,000 + \frac{N}{2} + \frac{Ny}{20} = 56\,250$$

and

$$N\left(\frac{10+y}{20}\right) = 31\,250$$

(d) *Answers*

When y is $3 per hour,

$$N = 48\,076 = 130 \text{ vehicles per day}$$

When y is $6 per hour,

$$N = 39\,062 = 110 \text{ vehicles per day}$$

In Example 19.1 it should be noted that the project would involve the taxpayers in an annual cost of $56 250 for construction and nett land acquisition less a maintenance saving of $25 000 p.a. – a nett cost of $31 250. The persons who gain are the road users; indeed the state itself would lose on such items as reduced income from petrol tax. However, the net income, or at least the net standard of living of the road users will increase, and, ultimately the domestic product, if the count is high. It is patently obvious that if the traffic count is high, the scheme should be adopted, but the problem of redistribution of wealth generated remains unsolved. The two social costs of drivers time might apply if the road had exclusively business or exclusively social use, since we all value our time differently during working and non-working hours.

In the example the social costs of land use were not considered. Land near the old (abandoned) route might well decrease in value, due to its likely increased distance from the markets, more than the land adjacent to the new route (which, being shorter, is likely to have a smaller catchment) would increase in value. The net charge, which will vary with land use, could be estimated and added to the costs in the equation, thereby raising N.

In this example of cost–benefit analysis we have committed the cardinal sin of considering net costs, which denies our definition that the B/C ratio is the ratio of total benefits divided by total costs. It is only permissible in this case since the critical B/C ratio is unity. For example, we should increase the land costs by not deducting the sale of the old land, and the income received from the latter should appear as a benefit. In this case, since the traffic warrant occurs when annual costs equal annual savings, the item would be self-cancelling as we would increase both sides of the equation by 5% of $375 000, the revenue from the sale of the land of the old road.

Had the analysis called for a B/C ratio of 2 – which could have been the case if we had an alternative project with such a value – then such short cuts should not be used. Even 15 km of maintenance *saving* should appear as a benefit, and 10 km of maintenance commitment as a cost; so should vehicle costs and human time costs.

As stated, in such studies justification is normally conceded when nett costs equal net savings, and, in the event, such studies can never be precise as the valuation of social time is indeterminate and it is known that shortening of a route always generates extra traffic (if travel costs reduce, travel demand increases!), but the amount of increase is not known.

It is interesting, in this connection, to note that the American Association of State Highway and Transport Officials recommends comparing a number of road projects using the definition that the

$$\text{B/C ratio} = \frac{\text{difference in road user costs}}{\text{difference in highway costs}}$$

the 'difference in road user costs' being savings.

This B/C *difference* ratio is the same as the B/C ratio only when it is unity, and it is incorrect to use it to compare schemes to give priority to ones with the highest difference ratio, tempting though it may be.

19.5   THE EFFECT OF TAX IN GOVERNMENT ANALYSIS

If a scheme is the concern of the entrepreneur, the costs incurred by him are those used in the analysis, e.g. if he uses fuel, the price used will include any fuel tax he has to pay. If, on the other hand, the scheme were governmental (a municipality normally pays fuel and similar taxes and is, in this case, in a different position to the state), then the government would not pay itself fuel tax, and this should not be included in the analysis. Neither would it include customs or import duty on imported items, since this would be analogous to robbing Peter to pay Paul. Thus all costs in government projects must be based on costs the state itself would incur, i.e. money it would actually pay out.

The question as to the inclusion of taxes on the income side is less apparent. If the government were prospecting for, then building and operating, an oilfield, it would not charge itself a fuel tax for construction fuel used on the cost side, and thus logically it should not credit itself with a tax income on the sales side. Indeed, for every barrel of oil produced, no *extra* tax income is generated, as each barrel sold merely replaces oil bought elsewhere, so no *extra* income should be credited. Thus no tax on output should be credited; indeed, were it so credited the state could make an unprofitable scheme profitable, merely by imposing a tax — a situation which is nonsensical. Shadow pricing would, however, be justified if the oil produced reduced imports or were for export.

19.6   THE PROBLEM OF SHADOW PRICING

It cannot be overstressed that shadow prices are not the concern of the entrepreneur who merely spends, and receives, his income in national currency. If his scheme involves foreign currency he will either be allowed to purchase it by the state, or not, in which case he cannot proceed.

However, in the case of governmental projects, shadow pricing must be applied if foreign currency is involved. To this end it must be dealt with like the other intangibles, i.e. social costs and benefits. If foreign goods or services are required, the costs so involved must be multiplied by the shadow prices. Conversely the income, from production generated by exports from the project, must be similarly treated.

Let us consider the simple case of a long-lasting irrigation scheme, the completion costs of which are $1 000 000, half of which (say for cement, machinery, and consultancies) is for imported articles, including $100 000 import duty, and half is for local goods and services. The market interest rate is 5% p.a. Operational costs, which are local, are $20 000 p.a., and all the products are to be exported for $60 000 p.a. There is a negative balance of payments and the shadow price is estimated at 1.5.

As the project is long term, investment need not be repaid, so only interest payments are concerned. To the entrepreneur, the annual costs are $70 000, being $50 000 interest on capital and $20 000 for operation. His income, from sales, is $60 000 p.a., and the B/C ratio is 0.87. Thus he would not proceed with the scheme, even if the foreign currency were available, unless a subsidy or grant-in-aid exceeding $10 000 were given annually.

As far as the government is concerned *economic* costs would be $500 000 local, plus $400 000 × 1.5 = $600 000 for imports (the $100 000 of tax is not paid). Thus annual (economic, though not financial) costs are 5% of $1 100 000, plus operating costs of $20 000, being $75 000.

The economic income is the product of the shadow price and the financial income, i.e. $60 000 × 1.5 or $90 000. The (economic) B/C ratio is now not the (financial) one of private enterprise but 90 000/75 000 or 1.2.

To the government, then, the scheme is viable, though it would not be to private enterprise. In the real world the government would also be able to borrow at less than the (private sector's) market interest rate. Let us assume it could borrow at 4% as against 5%. This would reduce annual economic costs of interest repayment by 20%, total annual costs reducing to $64 000, and the B/C ratio would increase to 1.41 — a sufficiently attractive figure to justify the scheme (unless there were a risk of a sizeable reduction in the overseas demand price for the product).

Digression: subsidies for export markets. This example illustrates that the scheme would be just viable for government which can borrow at 4% interest (B/C ratio unity) if their *economic* income were to drop to $64 000; that is, before shadow price adjustment, the financial sales income (in foreign currency) was only $42 670. This is due solely to the state's need to acquire foreign currency to balance its terms of trade, or to be able to import other foreign goods.

Consider, now, the case of private enterprise when annual sales are $60 000 (worth $90 000 to the state). Annual costs are $70 000, so an annual subsidy of at least $10 000 would be necessary to make the company viable, whilst one of $20 000 would enable it to show a profit. If government gave an annual subsidy of $20 000 it would gain the equivalent of $90 000 p.a. in foreign currency, and would tax the firm on its profits of $10 000; it would also tax the lender of capital on his annual interest of $50 000, and the workers on their salaries, and is thus likely to more than recoup its subsidy.

PROBLEMS

Problem P19.1

With the scheme in Problem P 18.3 it is estimated that other landowners using the new infrastructure can change from ranching to dairying, as dairy products can then be marketed. It is calculated that their net income will increase by $1 million p.a., and the tax rate is 50%. What is the effective B/C ratio of the scheme to government?

Problem P19.2

With the scheme in Problem P19.1, if the shadow price were 2, and half of the production were exported, what would be the economic B/C ratio to government if additional exporting transport cost $100 000 p.a. in foreign currency?

Problem P19.3

An existing road (scheme A) has had its costs written off, but vehicle and social costs are estimated at $2.5 million p.a. Two shorter routes are being considered, scheme B costing $500 000 p.a. for interest on construction costs, and $250 000 p.a. for interest on land acquisition, and scheme C, still shorter, costing $900 000 p.a. and $200 000 p.a. respectively. The vehicle and social costs would drop to $1.5 million p.a. for scheme B and to $1 million p.a. for scheme C. If either scheme B or C were built, the land of the existing road could be sold, and would result in an income of $100 000 p.a.

Calculate the B/C and differential B/C ratios of schemes B and C, and argue as to which scheme, if either, should be adopted.

Chapter 20

Private Sector Cost–Benefit Analysis and Subsidies

20.1 PUBLIC SECTOR JUSTIFICATION

In Chapter 19 it was illustrated that public sector schemes could often be justified where the private sector could not operate. In such cases the state—or municipality—can either accept the commitment, or subsidize private enterprise to do so. The usual means of such subsidy is by a 'grant-in-aid'. These grants may, on face value, be inequitable, since one industry or firm may receive them but not another, but can be justified in the public interest.

20.2 SCHEMES INVOLVING SOCIAL COSTS OR BENEFITS

Such cases apply where the scheme involves externalities which are of value, or negative value, to the public, which is not otherwise involved with the scheme. For example, a new dam and irrigation scheme, or an airfield, may involve a road detour, increasing the distance between towns or to farms lying beyond the scheme. In this case a social cost is involved, and some form of compensation may be required from the developer. The more usual case is that such a scheme may involve the construction of a new rail link, or a road, or the extension of an electricity network. In this case, the entrepreneur may object to paying for all, or any, of the cost.

Let us consider the case of an entrepreneur developing an agricultural scheme in unused land, with a market interest rate of 5%. Suppose that his anticipated sales income is $100 000 p.a. with running expenses for local labour at $60 000 p.a., and with the project involving $500 000 for land, dams, and general development, and $500 000 for road construction.

If he has to pay for the road, then the annual costs of the project (which can be regarded as everlasting) are $50 000, plus $60 000 for labour, and the B/C ratio is $100 000/110 000 = 0.91. He will not proceed. If, on the other hand, government were to pay for the road, his B/C ratio would increase to 1.18, and the scheme might prove viable. Quite often in such cases, government will offer to meet 50% of the infrastructure costs – in that case by making a grant-in-aid of $250 000 towards road construction. Thus the entrepreneur would then have to borrow or invest $750 000 – $37 500 p.a., which still gives a positive B/C ratio of 1.03.

The question is posed as to why government should give public money to the concern in question.

20.3 JUSTIFICATION OF GRANTS-IN-AID ON FINANCIAL AND POLITICAL GROUNDS

Suppose, in the last example, government decided to give a grant-in-aid of $500 000, or to build the road itself, which amounts to much the same thing.

Firstly let us consider costs. If private enterprise is working on a market rate of interest of 5%, government is likely to be able to borrow at say 4%. Thus, since the road is taken as everlasting (we have ignored maintenance and rehabilitation costs in this example) the annual cost to government is 4% of $500 000, i.e. $20 000. The nett profit generated by the scheme itself will be $100 000 less $60 000 for labour and less interest now of $25 000, that is $15 000. The tax rate on this will vary with the government – and from time to time, with its budget – but it is liable to be at least 50%, so $7500 may be anticipated in tax income to government.

Next we must consider the $60 000 paid out in wages. It is likely that the labour force will pay both direct (income) and indirect (sales) tax – again the extent will vary – but it may well average 25%, or $15 000 p.a., which is also an additional taxation income to government.

Thus the total tax annual income to government from the scheme will be of the order of $22 500, as against the cost of only $20 000, being the annual interest on the state-financed road grant-in-aid. Thus government will increase its annual income by $2500, and will have gained a road which itself will generate more income following use by others.

However, has the government increased its taxation revenue by encouraging the scheme, or has it merely attracted workers from one employer to another? In general, with the Malthusian growth of population, it is likely to have created new wealth, new employment; in other cases where population is static and employment is full, workers will only leave one employer to work for another if wages are higher, and the net gain to government would only be in extra tax on the extra profit.

However, there is a further indirect gain which might be classified as financial. The worker's extra wealth will be spent, which will generate more employment in commerce and supporting industry, which again will increase the government's taxable income.

It may be concluded that there are usually good political and financial reasons for subsidizing projects in private enterprise, particularly those which might have been viable in the public sector, but that it is difficult to quantify these reasons with any exactitude.

20.4 JUSTIFICATION OF GRANTS-IN-AID ON INTERNAL ECONOMIC CONSIDERATIONS

In Section 20.3 we restricted the analysis to possible financial returns, and a financial B/C analysis. In economic studies we must make allowance for social benefits and social costs. In the previous example we considered financial data from an agricultural scheme, but ignored the externalities. The new road will add to land values in the vicinity, even beyond the project, and land will become more productive in that the tertiary costs of transport of goods to the market will be reduced, and profits will increase.

Such profits will be to the account of the landowners concerned, but they, in turn, will pay more tax and increase government's income thereby. This, in return, will add to justification of a grant-in-aid.

In a B/C analysis, estimates must be made of the change of land use and hence of its value or earning capacity, and the amount added to benefits. However, it is not applicable to the entrepreneurial benefits, only to government's.

20.5 CONSIDERATIONS OF RAILWAY AND ELECTRICITY TARIFFS

Had it been decided to build a new railway line, rather than a road, the question is posed as to whether this would be to the account of the entrepreneur or to the state (assuming that the railways were state owned). The same applies to other facets of the infrastructure, such as power. This is due to the fact that whereas a road is of direct benefit to all, a railway may not be so to the same degree, and a high voltage power supply will be of no use to local farmers unless expensive transformer stations are built.

If the entrepreneur were required to finance the rail or power lines, special rates for the supply of the service would have to be negotiated, since normal tariffs include for the provision of a normal infrastructure by the authority concerned. On the other hand, similar considerations might apply in reverse were the entrepreneur not required to finance the development but merely to pay for the service provided, since the ratio of the service to infrastructure costs may be out of proportion to the national average.

Of course, there might have been just insufficient justification for those services before the development, which could then make them economically viable to other users. It can be concluded that each case must be treated on its merits, that no hard-and-fast rule can be laid down.

20.6 GRANTS-IN-AID TO EXISTING SCHEMES

Often commodity prices change and existing undertakings cease to be viable, that is their B/C ratio becomes less than unity, and grants-in-aid are sought. This applies equally to state-run nationalized industries. Typical cases are those of mines, where drops of 50% in ore prices are not unknown, and in the case of railways, where unrestricted road services attract trade and leave insufficient custom for the railways to remain viable with their reduced traffic.

Such cases are of equal concern to the engineer, and although no rigid rules can be applied, the same bases of methodology apply.

Questions to be asked are:

(a) Is the problem temporary; in future will a rise in commodity prices ensure a return to viability?
(b) Is the process obsolescent?
(c) Is alternative, preferably local, employment available?
(d) What is the magnitude of the loss?
(e) What would be the effect of closure on other dependent industries?
(f) What are the earnings of foreign currency?

If the loss is likely to be temporary, it is obvious that a subsidy is likely to be warranted, but an analysis should be undertaken with the amount of subsidy and the future likely benefits discounted to PDV with a series of risks.

If obsolescence is the problem, subsidy should only be considered if the work force will be unemployed, and the financial (if not social) cost of this unemployment entered as a debit. In this case the savings in not subsidizing must be measured against the alternative cost of unemployment.

The effect of closure in other industries must always be considered. The outstanding case is that of coal mines, as so many industries rely on a cheap source of coke. In this case the necessary subsidy must be measured against the increased costs for other industries if they have to obtain alternative supplies at higher costs.

Developing countries often have an economy based on the export of a single commodity: Zambia relies on copper, Mauritius on sugar. If the industry developed a negative B/C ratio and closed, the country concerned would be unable to earn foreign currency, and in the long run, would not be able to import foreign goods. Such countries either have to subsidize their basic industries, or devalue their currency. In the short run such devaluation would enable their exports to be sold competitively on world markets, and the income, that is benefits, in *local currency*, would rise, restoring the B/C ratio and viability. However, in the long run, since imports would cost more, costs would rise, and the B/C ratio would again drop. The amount of the drop will depend on many factors not connected with the industry, but the action will seldom result in a long-term solution.

Whilst the alteration of exchange rates is a complicated and problematical exercise, the B/C ratio of the project can be increased to unity with a grant-in-aid against the product of sales and shadow price as a benefit. This was illustrated in the Digression in Section 19.6.

PROBLEMS

Problem P20.1

A municipality owns roads in the central business district (CBD), where land values are $100/m^2, and the borrowing rate is 5% p.a. It proposes to convert some traffic lanes into kerbside parking sterilizing 15 m^2 per car, and to charge 5 cents/h, at which it expects to sell 1500 hours of parking p.a. It can hire meters at $25/yr each, and supervisors to police 250 cars per man at $5000 p.a. Road construction costs $10/m^2, and maintenance $0.5/(m^2/yr). Calculate: (1) the financial B/C ratio; (2) the economic B/C ratio; and (3) the hidden social subsidy from the ratepayer.

Problem P20.2

In a small country with negligible exports, ore is discovered. Consultants advise that with imported machinery costing $50 million, local materials and labour costing $60 million, and a 40 year life, ore could earn $8 million p.a. in foreign currency. However, local operating costs would be $1 million, and expatriate expertise $1 million p.a. also. The equipment manufacturers will loan money at 10% p.a., and local capital is available at 5%.

Calculate the subsidy necessary if the scheme is commissioned to gain foreign currency, and the shadow price to justify the scheme, ignoring indirect effects on the economy.

Chapter 21

Income Growth and Benefits

21.1 THE COMPLICATION

A complication arises in the calculation of the B/C ratio if the future income is liable to grow. If the increase is due to inflation, it should normally be ignored, but if it is due to an increasing demand, or to an inelastic supply, the calculations of PDV of income must be undertaken individually, year by year. Increasing (or reducing) sales price of the product would recur with a change in demand for the product.

If income growth, for any reason, is greater than the market interest rate, the PDV of each successive year's income will increase, and, sooner or later, benefits of *any* scheme will exceed costs, and the B/C ratio will become positive — but this may be so far off that its discounted effect may be negligible, in which case the project should be shelved until a future date. Obviously few entrepreneurs would be prepared to wait a decade or two to show profits.

21.2 THE CASE OF INCREASING PRICES

An example could be that of an oilfield development: the case of perfect competition and an inelastic supply, where sale prices of oil will, in the long run, increase due to demand exceeding supply.

Example 21.1 Benefit–cost ratio of a long-life oilfield

It will cost, on commissioning, $100 million to develop an oilfield. The market interest rate is 5%, and operational costs can be neglected for simplification. At a given constant output a long life is anticipated, and sales prices are estimated at $4 million after the first year, increasing, in terms of constant (PDV date) dollars, at 10% p.a. At the beginning of each subsequent year we have to calculate the B/C ratios of the project.

As the oilfield is long lasting, costs will thus approximate to the interest on capital, being $5 million p.a.

Table 21.1 can now be compiled, showing the annual income and the B/C ratio, in any year. It can be seen that the annual B/C ratio rises until the fourth

Table 21.1 Viability of an oilfield

End of year	Income ($ million) in past year	Expenditure ($ million) in past year	B/C ratio	PDV of income ($ million) in past year
1	4.0	5.0	0.80	3.81
2	4.40	5.0	0.88	3.99
3	4.84	5.0	0.97	4.18
4	5.32	5.0	1.06	4.38
5	5.86	5.0	1.17	4.59
6	6.44	5.0	1.29	4.81
7	7.09	5.0	1.42	5.04
8	7.79	5.0	1.56	5.28
9	8.57	5.0	1.71	5.53
n	$4(1.1)^{n-1}$	5.0	$\dfrac{4(1.1)^{n-1}}{5}$	$\dfrac{4(1.1)^{n-1}}{1.05^n}$

year — the first which is profitable — but, having shown a loss until then, the scheme is not yet viable after four years.

The total income, in constant dollars, discounted from year n to the commissioning (present day) date is

$$\sum_{1}^{n} \frac{4(1.1)^{n-1}}{1.05^n}$$

and this will be equal to $100 (we are dealing in $millions), when the B/C ratio first becomes unity. Thus

$$\sum_{1}^{n} \frac{4(1.1)^{n-1}}{1.05^n} = 100$$

from which we can calculate that $n \simeq 17$ years (Table D can be used to the limit of accuracy that is justified). Thus the scheme would have to operate for about 17 years before it became viable.

Example 21.1 Benefit–cost ratio of a short-life oilfield

Suppose, now, that the oilfield in Example 21.1 had a 10 year estimated life. We have to calculate the B/C ratio (which must be less than unity as life is less than 17 years).

Firstly Table 21.2 should be compiled giving income for each of the 10 years, and its discounted value. The B/C ratio is now $47.39/100 \simeq 0.47$ so the field is again proven to be not viable.

However, it could be made viable either by delaying its opening, or by reducing the output. If the project were delayed 10 years, the original sales price of oil would be $10.36, not $4 million, and the B/C ratio would become 1.22, not 0.47. Similarly, if output were reduced, more would be sold later, at higher prices, and again viability would increase, as the sales price is increasing faster than the market interest rate.

Table 21.2   Viability of a short-life oilfield

End of year	Income ($ millions)	PDV of income ($ millions)
1	4.00	3.81
2	4.40	3.99
3	4.84	4.18
4	5.32	4.38
5	5.86	4.59
6	6.44	4.81
7	7.09	5.04
8	7.79	5.28
9	8.57	5.53
10	9.42	5.78
Total		47.39

The PDV of any annual income from year n is the product of annual income and the increase in price, divided by the market interest rate factor, i.e.

$$\text{the given income} \times \left(\frac{1+R}{1+r} \right)^n$$

where R is the rate of increase of sales price, in *constant* dollars, and r is the market interest rate. Thus postponement of a project is economically justified when R is greater than r. Note that this does not apply with inflation where the value of money is not constant, but deflating.

21.3   THE CASE OF INCREASING DEMAND

In this case the income value per unit remains constant but the demand increases. The classic example of this is the case of a road, since in most countries the annual traffic count increases, or did before the energy crisis.

The same principles apply as in the previous case. It is conventional to design a road to last between 10 and 20 years before it needs reconstruction or rehabilitation. Whilst traffic has, in many countries, been increasing at 5% to 10% p.a., it is unlikely to do so with the increasing price of fuel, and long-term forecast growth rates exceeding 5% p.a. are open to question. Since this is normally less than the market interest rate, the PDV of future income (social savings) is unlikely to increase; so, although the B/C ratio will increase with the life expectancy, the discounted B/C ratio will not increase. This is the case of $R < r$.

Example 21.3 Justification of a new road

On completion a road will cost $ 50 000. The market interest rate is 6%, and traffic *savings* will average $ 7000 a year at the time of completion, and will increase at 5% p.a. for the design life of 10 years. We have to calculate the viability by means of a B/C ratio, assuming no residual value.

Table 21.3 Viability of a new road

End-of year	PDV of 'income' ($ thousands)	PDV of investment ($ thousands)
1	6.93	—
2	6.86	—
3	6.79	—
4	6.73	
5	6.66	—
6	6.59	—
7	6.53	—
8	6.46	—
9	6.40	—
10	6.34	—
Total	66.29	50.0

Since the annual 'income' will change we have to calculate the original PDV, not an annual figure. The PDV of 'income' earned in the nth year is $(1.05^n/1.06^n) \times \$7000$, which is approximately $\$7000/1.01^n$. We can tabulate the PDV of this income in Table 21.3.

From this it can be seen that the B/C ratio is 1.3 making it viable. Greater accuracy is not justified within the limits of accuracy of forecast data.

It cannot be overstressed that accuracy in such cases is not justified. If, for example, traffic in this example grew at 6%, the same as the interest rate, the growth would balance the discount figure and the B/C ratio would increase to 1.4, and who can forecast the traffic growth with great accuracy? In the real world, road schemes adjudged by an experienced engineer to be viable, are usually proved to be so. The problem then becomes not the justification of a scheme, but the assessment of its priority with limited financial and labour resources, the B/C ratio being needed to compare alternative schemes to justify priority. That being so, errors due to approximation, being relative, are even less significant.

PROBLEMS

Problem P21.1

It is proposed to develop a gravel pit for road construction purposes. The costs will be $\$1\,000\,000$ for the land, payable on purchase, with development charges of $\$500\,000$ after 6 months, and $\$500\,000$ on completion of development 6 months later. The reserves are estimated at $2\,000\,000$ m^3, and the planned output is $500\,000$ m^3 per year. Income will be at year-end. It is anticipated that the constant demand and reducing local reserves will permit of a price of $\$3$/m^3 for the first year, increasing at 15% yearly in terms of constant dollars. The market interest rate is 10%. If the land can be sold after use for $\$100\,000$, and annual year-end production costs are $\$300\,000$, determine the B/C ratio.

Problem P21.2

A factory installs plant, which costs $1 million payable on commissioning, to produce castings, the labour and materials costing $10 per casting. It expects to sell the plant after 5 years for $200 000, the market interest rate being 8%.

It anticipates a falling demand, and hence prices, for its products according to the following table:

Year	1	2	3	4	5
Number	8000	7000	6000	5000	4000
Average price ($)	80	70	60	55	50

Assume year-end accounting, and calculate the B/C ratio and the discounted profit.

Problem P21.3

If 10% p.a. inflation occurred with the project in Problem P21.2, discuss the change in profitability and the B/C ratio, with and without a change in the market interest rate.

Chapter 22

Initial Cost Maintenance, and Risk

22.1 THE IDEAL CHOICE OF INITIAL INVESTMENT

The main objective of the engineer should be the selection of his design parameters not to minimize initial cost, but to minimize the overall costs, that is to work to the lowest economic cost. Thus, *for a given service*, the sum of discounted construction and maintenance costs must be minimal. In other words his design parameters should be chosen by him such that his client's investment, both for the initial construction and for future maintenance, should be minimal.

In certain cases, the client may only be concerned with initial cost. This would be the case with, say, a car manufacturer who wishes to produce a car to sell at a lower cost than his rivals. He may choose to use fewer coats of paint to achieve such an aim, or to use a cheaper metal for his cylinder blocks, irrespective of the fact that his product may have only half the life, and will cost more to maintain. In this case the entrepreneur will play the role of the piper who calls the tune; he will be laying down the policy, the parameter of minimizing initial cost, to his engineer.

Usually, however, the engineer has a freer hand, in which case he should attempt to design and produce for the lowest economic cost. This applies particularly in public works, where the lowest initial price seldom results in the lowest costs.

22.2 THE PROBLEMS OF MAINTENANCE COSTS

Whereas, at the design stage, construction costs can be forecast with considerable accuracy, the same does not apply to maintenance. In the case of the car, how can one assess the increase in cost of repairs of reduction in life that a reduction of six coats of paint to four will bring about? This will certainly depend on how intensively the car is driven. For example, if a car is driven at 100 000 km/yr, the effect will be minimal, as mechanical wear is liable to be the cause of scrapping the car; whilst, if it is driven at only 10 000 km/yr, rust, rather than wear, will be the cause. Patently, too, the effects will be different in wet and dry conditions.

In all cases, however, some estimate of the amount of future maintenance, and its timing and cost, must be made, and discounted to PDV. This also raises the problems of inflation and of future interest rates.

As far as inflation is concerned, as has been stressed, this should be ignored, unless it is anticipated that a specific maintenance operation will escalate in price at a different rate to that of general inflation. In this case the cost of future operations should be based on the present price multiplied by the anticipated increase less the general inflation increase rate. Such a factor could, of course, be negative.

For example, if an inflation rate of 10% is anticipated, and the maintenance will consist exclusively of, say, bitumen which, due to inelastic supply, may inflate at 15%, it would be reasonable to increase the cost of future maintenance by 5% p.a. in the analysis. The inflation rate of 10% may generally be ignored, since productivity is likely to be constant, that is output per man-hour will be unchanged.

If it is difficult to forecast maintenance expenditure, it is equally difficult to forecast future inflation rates, and it may be equally difficult to forecast future market interest rates; so, normally, the interest rate prevailing when a project is commissioned is used. It is again stressed that great accuracy with future costs can never be achieved, so economic analysis merely serves to justify a project within broad limits, to establish priorities between schemes, and to justify design parameters.

22.3 THE OPTIMIZATION OF ECONOMIC COSTS

Two engineers, working on the same project, will seldom produce identical designs. What is important is that the economic costs should be approximately the same, that the product of initial cost and discounted maintenance should be reasonably similar with both solutions to the problems.

Let us consider a road design, with a market interest rate of 5%, and a 20 year life. Two solutions are offered: scheme A, costing $95 000, with strengthening costs of $20 000 estimated after 10 years; and scheme B, costing $80 000, with strengthening costs of $50 000 also after 10 years. The total costs, are $115 000 and $130 000 respectively, but we must discount the strengthening costs to PDV.

From Table B we see that we have to multiply future costs by 0.61391 to obtain the sum, which if invested, would grow sufficiently to pay for the strengthening. Thus we can see that scheme A will need an investment of $95 000 + $12 280 = $107 280, whilst scheme B will cost $80 000 + $30 700 = $110 700, and will thus be slightly more expensive, i.e. scheme A with the higher initial cost and the lower future cost will be the more economic. However, the saving is only of the order of 3%, which would change with future changes of interest, so for practical purposes, both designs would be acceptable.

Suppose now that the schemes would both last 20 years, with no strengthening costs, but with annual maintenance costs of $1000 and $2500 p.a. respectively. To follow the rules we use Table C to determine the PDV of maintenance. To obtain the PDV of maintenance expenditure we obtain a multiplication factor of 12.4622, from which we can determine costs as follows:

> Scheme A $95 000 + $12 500 = $107 500
>
> Scheme B $80 000 + $31 200 = $111 200

and again scheme A proves to be marginally cheaper.

If, now, as is possible, both schemes could be everlasting by excessive annual maintenance of $2500 and $3000 respectively, it would be simpler not to consider PDV but annual costs. In this case the construction cost needs never be repaid, only the 5% interest, so *annual* costs are:

> Scheme A $4750 + $2500 = $7250 p.a.
>
> Scheme B $4000 + $3000 = $7000 p.a.

In this case scheme B is the more economic.

It can be seen from these examples that there is no unique solution to the design of a road, which explains why so often, different consultants use completely different designs for similar problems, yet they could justify their work on economic, rather than on financial, or initial cost analysis.

Indeed, road transportation studies indicate that the more that is spent on roads, the lower is the expenditure on the operation of vehicles. If total costs are summed, a U-shaped curve with a flat bottom can be deduced, so the choice of expenditure (i.e. design choice) on road construction can vary considerably — at the expense of the operator—with only marginal changes to overall transportation costs. This is true in many other situations and explains the wide range of satisfactory design solutions in many engineering operations, where structural failure is not liable to be catastrophic.

22.4   DESIGN RISK

According to statistical theory a project must be infinitely overdesigned to ensure almost zero risk, so it is patently obvious that risk must be evaluated against cost, and a decision made as to what risk to accept.

For example, if a river is known to flood when the annual rainfall exceeds the mean, the higher (and stronger) that protection works are designed, the greater the cost, and the less will be the risk of flooding. Such flooding may result in damage to buildings, the infrastructure, or merely in the loss of a crop. The damage must be quantified in financial terms, and plotted against the construction (and discounted maintenance!) costs.

Patently when risk may involve human life, only a low risk can be taken; this is why, in structural design, permissible design stresses are considerably less than field test data. Conversely, in many engineering works, such as roads, failure can only be measured in financial terms, so higher risks are justified.

Example 21.1

A typical example would be the overlay for strengthening of an old road. Let us

Table 22.1 Overlay costs and risk on an old road

$\bar{x}$	Standard deviation (SD)	Risk of failure (%)	Design deflection (mm)	Thickness* (mm)	Cost* ($1000/km)
0.5	0.2	$50(\bar{x})$	0.50	90	27
0.5	0.2	$16(\bar{x}+SD)$	0.70	130	39
0.5	0.2	$7(\bar{x}+1.5SD)$	0.80	140	42
0.5	0 2	$5(\bar{x}+1.65SD)$	0.83	145	43.5
0.5	0.2	$2.5(\bar{x}+2SD)$	0.90	150	45

consider such a road. Deflection surveys show an average deflection of 0.50 mm with a standard deviation of 0.2 mm. For any given traffic and deflection the required overlay thickness (and hence cost) is known, and if a normal Gaussian distribution is assumed, the risk can be calculated. Table 22.1 lists the data, which are plotted on Figure 22.1.

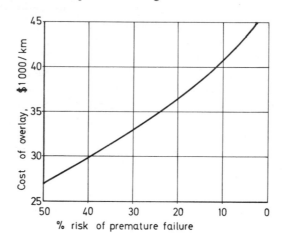

Figure 22.1   Cost of road strengthening *versus* risk

It can be seen that any sum between $27 000/km and $45 000/km can be spent for a risk of premature failure of between 50% and 2.5%. It should be noted that, theoretically, infinite thickness and cost are necessary for no risk, so there is an indefinite choice of design parameters.

The design engineer must therefore make a decision based on economics and risk that he, or his client, should take. In this case, 'failure' will merely constitute a loss of ride, and is unlikely to be catastrophic; moreover, other indeterminate factors, such as future traffic volumes and climate, must be considered. If funds are available but future facilities for overlays are problematical, a 2.5% risk might be

* It is assumed that overlay thickness and cost is known for all deflections.

reasonable, whilst if funds are in short supply, he might opt for a 25% risk. If $90 000 were available, he could strengthen 2 km, with a 2.5% risk, or 2.6 km with a 25% risk.

In space exploration, all components are designed to perform with a preselected risk of failure. If this were not so, designs would be so expensive as to prohibit all space research; rockets would possibly even be too heavy to lift off!

22.5 THE SELECTION OF DESIGN RISK

This is an area which relies upon the engineer's judgement, but nevertheless, if projects are to be undertaken, judgement must be quantified.

Risk is a problem for statistics, and Gaussian (normal) distributions are usually assumed for mathematical simplicity, although in many cases they may not be exact, or even justified. For example, if a certain high density is called for in soil compaction the distribution of test data is likely to be skewed, as it is easier to obtain less, than more, compaction. However, a normal distribution can be used and the risk of excessive premature settlement, i.e. deformation, calculated; on this figure, a decision as to whether to accept, reject, or conditionally accept (for reduced payment) can be taken, when a contractor's compaction density is below specification.

The degree of acceptable risk will vary with the implications. If it is a question of the life of a road, or of a sewer, or of a car, or a lathe, a high degree can be accepted; whereas, if it is a question of a bridge, or a jet engine, a lower degree must be chosen, which will result in cost escalation.

The greatest problems occur when human life is concerned. Many moralists consider the value of human life to be infinite; if this view were accepted, then no high speed car or road should be designed, no aircraft, and indeed, since statistics of fatal accidents show that loss of life in construction increases with the magnitude of a project, few projects should be undertaken. Thus it follows that life must be quantified in economic terms: a guide to such valuation would be the loss in discounted potential earning capacity of the persons likely to be involved. Actuaries are used to quantifying the value of human life, and there is no need for engineers to baulk at so doing.

PROBLEMS

Problem P22.1

An airport is being planned for a city. The cost of each runway is $10 000 000. From a study of windrose diagrams it is estimated that with a single runway layout, 97% serviceability will result, whilst a double runway layout will give 99% serviceability. When the airfield is closed, the following costs are estimated, per aircraft movement, on diversion:

Use of alternative airport and ground transport	$2000
Social costs, passengers time	$1000
Social costs, general inconvenience	$1000

Assuming that runways are designed for a 10 year life, and that the interest rate is 5% p.a., the number of movements per day must be calculated at which the choice of the single or double runway layout is marginal, assuming that the airfield were governmentally financed.

Problem P22.2

An isolated factory is dependent upon its power supply. From past experience it has been determined that there will be a 0.1% risk of failure, and that the average resulting shutdown period is 1 day. A 1 day shutdown will result in a loss of profits of $4000. A standby plant costing $15 000 is being considered at a market investment rate of 6% p.a.
Determine whether the standby plant should be purchased.

Problem P22.3

A dyke is proposed for river protection. The higher the dyke the greater the costs, and the lower the risk of flooding. Estimated data are indicated in the following table:

Height of dyke (m)	2	3	4	5	6	7
Cost of dyke ($1000)	10	25	40	65	100	150
Risk of flooding (times a year)	3	2	1	0.5	0.1	0.05

If the damage by flooding is estimated at $10 000 each time it occurs, what design height should be selected if money can be borrowed at: (a) 10%; and (b) 20%?

Interest Tables

TABLE A TERMINAL VALUE OF A SINGLE SUM AT COMPOUND INTEREST

The amount to which \$1, invested today, will increase in n years with interest rate r per annum $= (1 + r)^n$

Interest % $(= 100r)$

n (years)	1.0	1.5	2.0	2.5	3.0	3.5	4.0	4.5	5.0	5.5
1	1.0100	1.0150	1.0200	1.0250	1.0300	1.0350	1.0400	1.0450	1.0500	1.0550
2	1.0201	1.0302	1.0404	1.0506	1.0609	1.0712	1.0816	1.0920	1.1025	1.1130
3	1.0303	1.0457	1.0612	1.0769	1.0927	1.1087	1.1249	1.1412	1.1576	1.1742
4	1.0406	1.0614	1.0824	1.1038	1.1255	1.1475	1.1699	1.1925	1.2155	1.2388
5	1.0510	1.0773	1.1041	1.1314	1.1593	1.1877	1.2167	1.2462	1.2763	1.3070
6	1.0615	1.0934	1.1262	1.1597	1.1941	1.2293	1.2653	1.3023	1.3401	1.3788
7	1.0721	1.1098	1.1487	1.1887	1.2299	1.2723	1.3159	1.3609	1.4071	1.4547
8	1.0829	1.1265	1.1717	1.2184	1.2668	1.3168	1.3686	1.4221	1.4775	1.5347
9	1.0937	1.1434	1.1951	1.2489	1.3048	1.3629	1.4233	1.4861	1.5513	1.6191
10	1.1046	1.1605	1.2190	1.2801	1.3439	1.4106	1.4802	1.5530	1.6289	1.7081
11	1.1157	1.1779	1.2434	1.3121	1.3842	1.4600	1.5395	1.6229	1.7103	1.8021
12	1.1268	1.1956	1.2682	1.3449	1.4258	1.5111	1.6010	1.6959	1.7959	1.9012
13	1.1381	1.2136	1.2936	1.3785	1.4685	1.5640	1.6651	1.7722	1.8856	2.0058
14	1.1495	1.2318	1.3195	1.4130	1.5126	1.6187	1.7317	1.8519	1.9799	2.1161
15	1.1610	1.2502	1.3459	1.4483	1.5580	1.6753	1.8009	1.9353	2.0789	2.2325
16	1.1726	1.2690	1.3728	1.4845	1.6047	1.7340	1.8730	2.0224	2.1829	2.3553
17	1.1843	1.2880	1.4002	1.5216	1.6528	1.7947	1.9479	2.1134	2.2920	2.4848
18	1.1961	1.3073	1.4282	1.5597	1.7024	1.8575	2.0258	2.2085	2.4066	2.6215
19	1.2081	1.3270	1.4568	1.5986	1.7535	1.9225	2.1068	2.3079	2.5269	2.7656
20	1.2202	1.3469	1.4859	1.6386	1.8061	1.9898	2.1911	2.4117	2.6533	2.9178
25	1.2824	1.4509	1.6406	1.8539	2.0938	2.3632	2.6658	3.0054	3.3864	3.8134
30	1.3478	1.5631	1.8114	2.0976	2.4273	2.8068	3.2434	3.7453	4.3219	4.9840
35	1.4166	1.6839	1.9999	2.3732	2.8139	3.3336	3.9461	4.6673	5.5160	6.5138
40	1.4889	1.8140	2.2080	2.6851	3.2620	3.9593	4.8010	5.8164	7.0400	8.5133
45	1.5648	1.9542	2.4379	3.0379	3.7816	4.7024	5.8412	7.2482	8.9850	1.1127
50	1.6446	2.1052	2.6916	3.4371	4.3839	5.5849	7.1067	9.0326	11.467	14.542
55	1.7286	2.2679	2.9717	3.8888	5.0821	6.6331	8.6464	11.256	14.636	19.006
60	1.8167	2.4432	3.2810	4.3998	5.8916	7.8781	10.519	14.027	18.679	24.840

(Contd.)

Table A (*Contd.*)

Interest % (= 100*r*)

n (years)	6.0	6.5	7.0	7.5	8.0	9.0	10.0	12.0	15.0	20.0
1	1.0600	1.0650	1.0700	1.0750	1.0800	1.0900	1.1000	1.1200	1.1500	1.2000
2	1.1236	1.1342	1.1449	1.1556	1.1664	1.1881	1.2100	1.2544	1.3225	1.4400
3	1.1910	1.2079	1.2250	1.2423	1.2597	1.2950	1.3310	1.4049	1.5209	1.7280
4	1.2625	1.2865	1.3108	1.3355	1.3605	1.4116	1.4641	1.5735	1.7490	2.0736
5	1.3382	1.3701	1.4026	1.4356	1.4693	1.5386	1.6105	1.7623	2.0114	2.4883
6	1.4185	1.4591	1.5007	1.5433	1.5869	1.6771	1.7716	1.9738	2.3131	2.9860
7	1.5036	1.5540	1.6058	1.6590	1.7138	1.8280	1.9487	2.2107	2.6600	3.5832
8	1.5938	1.6450	1.7182	1.7835	1.8509	1.9926	2.1436	2.4760	3.0590	4.2998
9	1.6895	1.7626	1.8385	1.9172	1.9990	2.1719	2.3579	2.7731	3.5179	5.1598
10	1.7908	1.8771	1.9672	2.0610	2.1589	2.3674	2.5937	3.1058	4.0456	6.1917
11	1.8983	1.9992	2.1049	2.2156	2.3316	2.5804	2.8531	3.4785	4.6524	7.4301
12	2.0122	2.1291	2.2522	2.3818	2.5182	2.8127	3.1384	3.8960	5.3502	8.9161
13	2.1329	2.2675	2.4098	2.5604	2.7196	3.0658	3.4523	4.3635	6.1528	10.699
14	2.2609	2.4149	2.5785	2.7524	2.9372	3.3417	3.7975	4.8871	7.0757	12.839
15	2.3966	2.5718	2.7590	2.9589	3.1722	3.6425	4.1772	5.4736	8.1371	15.407
16	2.5404	2.7390	2.9522	3.1808	3.4259	3.9703	4.5950	6.1304	9.3576	18.488
17	2.6928	2.9170	3.1588	3.4194	3.7000	4.3276	5.0545	6.8660	10.761	22.186
18	2.8543	3.1067	3.3799	3.6758	3.9960	4.7171	5.5599	7.6900	12.375	26.623
19	3.0256	3.3086	3.6165	3.9515	4.3157	5.1417	6.1159	8.6128	14.232	31.948
20	3.2071	3.5236	3.8697	4.2479	4.6610	5.6044	6.7275	9.6463	16.367	38.338
25	4.2919	4.8277	5.4274	6.0983	6.8485	8.6231	10.835	17.000	32.919	95.396
30	5.7435	6.6144	7.6123	8.7550	10.063	13.268	17.449	29.960	66.212	237.38
35	7.6861	9.0623	10.677	12.569	14.785	20.414	28.102	52.800	133.18	590.67
40	10.286	12.416	14.974	18.044	21.725	31.409	45.259	93.051	267.86	1469.8
45	13.765	17.011	21.002	25.905	31.920	48.327	72.890	163.99	538.77	3657.3
50	18.420	23.307	29.457	37.190	46.902	74.358	117.39	289.00	1083.7	9100.4
55	24.650	31.932	41.315	53.391	68.914	114.41	189.06	509.32	2179.7	22644
60	32.988	43.750	57.946	76.649	101.26	176.03	304.50	897.59	4384.1	56346

Interest tables given here are based on the data of, and reproduced by permission of, the Institution of Civil Engineers, London.

TABLE B   PRESENT VALUE OF A SINGLE SUM

The present day value (PDV) of \$1 n years hence, when discounted at interest rate r per annum $= (1 + r)^{-n}$

Interest % $(= 100r)$

n (years)	1	1.5	2	2.5	3	3.5	4	4.5	5	5.5
1	0.99010	0.98522	0.98039	0.97561	0.97087	0.96618	0.96154	0.95694	0.95238	0.94787
2	0.98030	0.97066	0.96117	0.95181	0.94260	0.93351	0.92456	0.91573	0.90703	0.89845
3	0.97059	0.95632	0.94232	0.92860	0.91514	0.90194	0.88900	0.87630	0.86384	0.85161
4	0.96098	0.94218	0.92385	0.90595	0.88849	0.87144	0.85480	0.83856	0.82270	0.80722
5	0.95147	0.92826	0.90573	0.88385	0.86261	0.84197	0.82193	0.80245	0.78353	0.76513
6	0.94205	0.91454	0.88797	0.86230	0.83748	0.81350	0.79031	0.76790	0.74622	0.72525
7	0.93272	0.90103	0.87056	0.84127	0.81309	0.78599	0.75992	0.73483	0.71068	0.68744
8	0.92348	0.88771	0.85349	0.82075	0.78941	0.75941	0.73069	0.70319	0.67684	0.65160
9	0.91434	0.87459	0.83676	0.80073	0.76642	0.73373	0.70259	0.67290	0.64461	0.61763
10	0.90529	0.86167	0.82035	0.78120	0.74409	0.70892	0.67556	0.64393	0.61391	0.58543
11	0.89632	0.84893	0.80426	0.76214	0.72242	0.68495	0.64958	0.61620	0.58468	0.55491
12	0.88745	0.83639	0.78849	0.74356	0.70138	0.66178	0.62460	0.58966	0.55684	0.52598
13	0.87866	0.82403	0.77303	0.72542	0.68095	0.63940	0.60057	0.56427	0.53032	0.49856
14	0.86996	0.81185	0.75788	0.70773	0.66112	0.61778	0.57748	0.53997	0.50507	0.47257
15	0.86135	0.79985	0.74301	0.69047	0.64186	0.59689	0.55526	0.51672	0.48102	0.44793
16	0.85282	0.78803	0.72845	0.67363	0.62317	0.57671	0.53391	0.49447	0.45811	0.42458
17	0.84438	0.77637	0.71416	0.65720	0.60502	0.55720	0.51337	0.47318	0.43630	0.40245
18	0.83602	0.76491	0.70016	0.64117	0.58739	0.53836	0.49363	0.45280	0.41552	0.38147
19	0.82774	0.75361	0.68643	0.62553	0.57029	0.52016	0.47464	0.43330	0.39573	0.36158
20	0.81954	0.74247	0.67297	0.61027	0.55368	0.50257	0.45639	0.41464	0.37689	0.34273
25	0.77977	0.68921	0.60953	0.53939	0.47761	0.42315	0.37512	0.33273	0.29530	0.26223
30	0.74192	0.63976	0.55207	0.47674	0.41199	0.35628	0.30832	0.26700	0.23138	0.20064
35	0.70591	0.59387	0.50003	0.42137	0.35538	0.29998	0.25342	0.21425	0.18129	0.15352
40	0.67165	0.55126	0.45289	0.37243	0.30656	0.25257	0.20829	0.17193	0.14205	0.11746
45	0.63905	0.51171	0.41020	0.32917	0.26444	0.21266	0.17120	0.13796	0.11130	0.08988
50	0.60804	0.47500	0.37153	0.29094	0.22811	0.17905	0.14071	0.11071	0.08720	0.06877
55	0.57853	0.44093	0.33650	0.25715	0.19677	0.15076	0.11566	0.08884	0.06833	0.05262
60	0.55045	0.40930	0.30478	0.22728	0.16973	0.12693	0.09506	0.07129	0.05354	0.04026

Interest % $(= 100r)$

n (years)	6	6.5	7	7.5	8	9	10	12	15	20
1	0.94340	0.93897	0.93458	0.93023	0.92593	0.91743	0.90909	0.89286	0.86957	0.83333
2	0.89000	0.88166	0.87344	0.86533	0.85734	0.84168	0.82645	0.79719	0.75614	0.69444
3	0.83962	0.82785	0.81630	0.80496	0.79383	0.77218	0.75131	0.71178	0.65752	0.57870
4	0.79209	0.77732	0.76290	0.74480	0.73503	0.70843	0.68301	0.63552	0.57175	0.48225
5	0.74726	0.72988	0.71299	0.69656	0.68058	0.64993	0.62092	0.56743	0.49718	0.40188

(Contd.)

Table **B** (*Contd.*)

n (years)	6	6.5	7	7.5	8	9	10	12	15	20
6	0.70496	0.68533	0.66634	0.64796	0.63017	0.59627	0.56447	0.50663	0.43233	0.33490
7	0.66506	0.64351	0.62275	0.60275	0.58349	0.54703	0.51316	0.45235	0.37594	0.27908
8	0.62741	0.60423	0.58201	0.56070	0.54027	0.50187	0.46651	0.40388	0.32690	0.23257
9	0.59190	0.56735	0.54393	0.52158	0.50025	0.46043	0.42410	0.36061	0.28426	0.19381
10	0.55839	0l53273	0.50835	0.48519	0.46319	0.42241	0.38554	0.32197	0.24718	0.16151
11	0.52679	0.50021	0.47509	0.45134	0.42888	0.38753	0.35049	0.28748	0.21494	0.13459
12	0.49697	0.46968	0.44401	0.41985	0.39711	0.35553	0.31863	0.25668	0.18691	0.11216
13	0.46884	0.44102	0.41496	0.39056	0.36770	0.32618	0.28966	0.22917	0.16253	0.09346
14	0.44230	0.41410	0.38782	0.36331	0.34046	0.29925	0.26333	0.20462	0.14133	0.07789
15	0.41727	0.38883	0.36245	0.33797	0.31524	0.27454	0.23939	0.18270	0.12289	0.06491
16	0.39365	0.36510	0.33873	0.31439	0.29189	0.25187	0.21763	0.16312	0.10686	0.05409
17	0.37136	0.34281	0.31657	0.29245	0.27027	0.23107	0.19784	0.14564	0.09293	0.04507
18	0.35034	0.32189	0.29586	0.27205	0.25025	0.21199	0.17986	0.13004	0.08081	0.03756
19	0.33051	0.30224	0.27651	0.25307	0.23171	0.19449	0.16351	0.11611	0.07027	0.03130
20	0.31180	0.28380	0.25842	0.23541	0.21455	0.17843	0.14864	0.10367	0.06110	0.02608
25	0.23300	0.20714	0.18425	0.16398	0.14602	0.11597	0.09230	0.05882	0.03038	0.01048
30	0.17411	0.15119	0.13137	0.11422	0.09938	0.07537	0.05731	0.03338	0.01510	0.00421
35	0.13011	0.11035	0.09366	0.07956	0.06763	0.04899	0.03558	0.01894	0.00751	0.00169
40	0.09722	0.08054	0.06678	0.05542	0.04603	0.03184	0.02209	0.01075	0.00373	0.00068
45	0.07265	0.05879	0.04761	0.03860	0.03133	0.02069	0.01372	0.00610	0.00186	0.00027
50	0.05429	0.04291	0.03395	0.02689	0.02132	0.01345	0.00852	0.00346	0.00092	0.00011
55	0.04057	0.03132	0.02420	0.01873	0.01451	0.00874	0.00529	0.00196	0.00044	0.00004
60	0.03031	0.02286	0.01726	0.01305	0.00988	0.00568	0.00328	0.00111	0.00023	0.00002

TABLE C   PRESENT VALUE OF AN ANNUITY

The present day value (PDV) of $1 per annum for n years when discounted at interest rate r per annum $= [(1 - (1 + r)^{-n}]/r$

AMORTIZATION VALUES

The reciprocal of equal end-of-year payments to redeem a loan of $1 at the end of n years and to provide interest on the outstanding balance at interest rate r per annum

Interest % ($= 100r$)

n (years)	1	1.5	2	2.5	3	3.5	4	4.5	5	5.5
1	0.9901	0.9852	0.9804	0.9756	0.9709	0.9662	0.9615	0.9569	0.9524	0.9479
2	1.9704	1.9559	1.9416	1.9274	1.9135	1.8997	1.8861	1.8727	1.8594	1.8463
3	2.9410	2.9122	2.8839	2.8560	2.8286	2.8016	2.7751	2.7490	2.7232	2.6979
4	3.9020	3.8544	3.8077	3.7620	3.7171	3.6731	3.6299	3.5875	3.5460	3.5052
5	4.8534	4.7826	4.7135	4.6458	4.5797	4.5151	4.4518	4.3900	4.3295	4.2703
6	5.7955	5.6972	5.6014	5.5081	5.4172	5.3286	5.2421	5.1579	5.0757	4.9955
7	6.7282	6.5982	6.4720	6.3494	6.2303	6.1145	6.0021	5.8927	5.7864	5.6830
8	7.6517	7.4859	7.3255	7.1701	7.0197	6.8740	6.7327	6.5959	6.4632	6.3346
9	8.5660	8.3605	8.1622	7.9709	7.7861	7.6077	7.4353	7.2688	7.1078	6.9522
10	9.4713	9.2222	8.9826	8.7521	8.5302	8.3166	8.1109	7.9127	7.7217	7.5376
11	10.3676	10.0711	9.7868	9.5142	9.2526	9.0015	8.7605	8.5289	8.3064	8.0925
12	11.2551	10.9075	10.5753	10.2578	9.9540	9.6633	9.3851	9.1186	8.8633	8.6185
13	12.1337	11.7315	11.3484	10.9832	10.6350	10.3027	9.9856	9.6829	9.3936	9.1171
14	13.0037	12.5434	12.1062	11.6909	11.2961	10.9205	10.5631	10.2228	9.8986	9.5896
15	13.8650	13.3432	12.8493	12.3814	11.9379	11.5174	11.1184	10.7395	10.3797	10.0376
16	14.7179	14.1313	13.5777	13.0550	12.5611	12.0941	11.6523	11.2340	10.8378	10.4622
17	15.5622	14.9076	14.2919	13.7122	13.1661	12.6513	12.1657	11.7072	11.2741	10.8646
18	16.3983	15.6725	14.9920	14.3534	13.7535	13.1897	12.6593	12.1600	11.6896	11.2461
19	17.2260	16.4262	15.6785	14.9789	14.3238	13.7098	13.1339	12.5933	12.0853	11.6077
20	18.0455	17.1686	16.3514	15.5892	14.8775	14.2124	13.5903	13.0079	12.4622	11.9504
25	22.0231	20.7196	19.5234	18.4244	17.4131	16.4815	15.6221	14.8282	14.0939	13.4139
30	25.8077	24.0158	22.3964	20.9303	19.6004	18.3920	17.2920	16.2889	15.3725	14.5337
35	29.4086	27.0756	24.9986	23.1452	21.4872	20.0007	18.6646	17.4610	16.3742	15.3906
40	32.8347	29.9158	27.3555	25.1028	23.1148	21.3551	19.7928	18.4016	17.1591	16.0461
45	36.0945	32.5523	29.4902	26.8330	24.5187	22.4954	20.7200	19.1563	17.7741	16.5477
50	39.1961	34.9997	31.4236	28.3623	25.7298	23.4556	21.4822	19.7620	18.2559	16.9315
55	42.1472	37.2715	33.1748	29.7140	26.7744	24.2641	22.1086	20.2480	18.6335	17.2252
60	44.9550	39.3803	34.7609	30.9087	27.6756	24.9447	22.6235	20.6380	18.9293	17.4500

Interest % ($= 100r$)

n (years)	6	6.5	7	7.5	8	9	10	12	15	20
1	0.9434	0.9390	0.9346	0.9302	0.9259	0.9174	0.9091	0.8929	0.8696	0.8333
2	1.8334	1.8206	1.8080	1.7956	1.7833	1.7591	1.7355	1.6901	1.6257	1.5278
3	2.6730	2.6485	2.6243	2.6005	2.5771	2.5313	2.4869	2.4018	2.2832	2.1065
4	3.4651	3.4258	3.3872	3.3493	3.3121	3.2397	3.1699	3.0373	2.8550	2.5887
5	4.2124	4.1557	4.1002	4.0459	3.9927	3.8897	3.7908	3.6048	3.3522	2.9906

(Contd.)

Table C (*Contd.*)

n (years)	6	6.5	7	7.5	8	9	10	12	15	20
6	4.9173	4.8410	4.7665	4.6938	4.6229	4.4859	4.3553	4.1114	3.7845	3.3255
7	5.5824	5.4845	5.3893	5.2966	5.2064	5.0330	4.8684	4.5638	4.1604	3.6046
8	6.2098	6.0888	5.9713	5.8573	5.7466	5.5348	5.3349	4.9676	4.4873	3.8372
9	6.8017	6.6561	6.5152	6.3789	6.2469	5.9952	5.7590	5.3282	4.7716	4.0310
10	7.3601	7.1888	7.0236	6.8641	6.7101	6.4177	6.1446	5.6502	5.0188	4.1925
11	7.8869	7.6890	7.4987	7.3154	7.1390	6.8052	6.4951	5.9377	5.2337	4.3271
12	8.3838	8.1587	7.9427	7.7353	7.5361	7.1607	6.8137	6.1944	5.4206	4.4392
13	8.8527	8.5997	8.3577	8.1258	7.9038	7.4869	7.1034	6.4235	5.5831	4.5327
14	9.2950	9.0138	8.7455	8.4892	8.2442	7.7862	7.3667	6.6282	5.7245	4.6106
15	9.7122	9.4027	9.1079	8.8271	8.5595	8.0607	7.6061	6.8109	5.8474	4.6755
16	10.1059	9.7678	9.4466	9.1415	8.8514	8.3126	7.8237	6.9740	5.9542	4.7296
17	10.4773	10.1106	9.7632	9.4340	9.1216	8.5436	8.0216	7.1196	6.0472	4.7746
18	10.8276	10.4325	10.0591	9.7060	9.3719	8.7556	8.2014	7.2497	6.1280	4.8122
19	11.1581	10.7347	10.3356	9.9591	9.6036	8.9501	8.3649	7.3658	6.1982	4.8435
20	11.4699	11.0185	10.5940	10.1945	9.8181	9.1285	8.5136	7.4694	6.2593	4.8696
25	12.7834	12.1979	11.6536	11.1469	10.6748	9.8226	9.0770	7.8431	6.4641	4.9476
30	13.7648	13.0587	12.4090	11.8104	11.2578	10.2737	9.4269	8.0552	6.5660	4.9789
35	14.4982	13.6870	12.9477	12.2725	11.6546	10.5668	9.6442	8.1755	66166	4.9915
40	15.0463	14.1455	13.3317	12.5944	11.9246	10.7574	9.7791	8.2438	6.6418	4.9966
45	15.4558	14.4802	13.6055	12.8136	12.1084	10.8812	9.8628	8.2825	6.6543	4.9986
50	15.7619	14.7245	13.8007	12.9748	12.2335	10.9617	9.9148	8.3045	6.6605	4.9995
55	15.9905	14.9028	13.9400	13.0836	12.3186					
60	16.1614	15.0330	14.0392	13.1594	12.3766					

TABLE D   SINKING FUND

The equal amount per annum invested after each of n years at interest rate r to accumulate $= r/[(1 + r)^n - 1]$

EQUAL ANNUAL INVESTMENT

The reciprocal of the amount to which equal end-of-year payment of $1 per annum will accumulate when invested for n years at an interest rate of r per annum

Interest % $(= 100r)$

n (years)	1	1.5	2	2.5	3	3.5	4	4.5	5	5.5
1	1.00000	1.00000	1.00000	1.00000	1.00000	1.00000	1.00000	1.00000	1.00000	1.00000
2	0.49751	0.49628	0.49505	0.49383	0.49261	0.49140	0.49020	0.48900	0.48780	0.48662
3	0.33002	0.32838	0.32675	0.32514	0.32353	0.32193	0.32035	0.31877	0.31721	0.31565
4	0.24628	0.24444	0.24262	0.24082	0.23903	0.23725	0.23549	0.23374	0.23201	0.23029
5	0.19604	0.19409	0.19216	0.19025	0.18835	0.18648	0.18463	0.18279	0.18097	0.17918
6	0.16255	0.16053	0.15853	0.15655	0.15460	0.15267	0.15076	0.14888	0.14702	0.14518
7	0.13863	0.13656	0.13451	0.13250	0.13051	0.12854	0.12661	0.12470	0.12282	0.12096
8	0.12069	0.11858	0.11651	0.11447	0.11246	0.11048	0.10853	0.10661	0.10472	0.10286
9	0.10674	0.10461	0.10252	0.10046	0.09843	0.09645	0.09449	0.09257	0.09069	0.08884
10	0.09558	0.09343	0.09133	0.08926	0.08723	0.08524	0.08329	0.08138	0.07950	0.07767
11	0.08645	0.08429	0.08218	0.08011	0.07808	0.07609	0.07415	0.07225	0.07039	0.06857
12	0.07885	0.07668	0.07456	0.07249	0.07046	0.06848	0.06655	0.06467	0.06283	0.06103
13	0.07241	0.07024	0.06812	0.06605	0.06403	0.06206	0.06014	0.5828	0.05646	0.05468
14	0.06690	0.06472	0.06260	0.06054	0.05853	0.05657	0.05467	0.05282	0.05102	0.04928
15	0.06212	0.05994	0.05783	0.05577	0.05377	0.05183	0.04994	0.04811	0.04634	0.04463
16	0.05794	0.05577	0.05365	0.05160	0.04961	0.04768	0.04582	0.04402	0.04227	0.04058
17	0.05426	0.05208	0.04997	0.04793	0.04595	0.04404	0.04220	0.04042	0.03870	0.03704
18	0.05098	0.04881	0.04670	0.04467	0.04271	0.04082	0.03899	0.03724	0.03555	0.03392
19	0.04805	0.04588	0.04378	0.04176	0.03981	0.03794	0.03614	0.03441	0.03275	0.03115
20	0.04542	0.04325	0.04116	0.03915	0.03722	0.03536	0.03358	0.03188	0.03024	0.02868
25	0.03541	0.03326	0.03122	0.02928	0.02743	0.02567	0.02401	0.02244	0.02095	0.01955
30	0.02875	0.02664	0.02465	0.02278	0.02102	0.01937	0.01783	0.01639	0.01505	0.01381
35	0.02400	0.02193	0.02000	0.01821	0.01654	0.01500	0.01358	0.01227	0.01107	0.00997
40	0.02046	0.01843	0.01656	0.01484	0.01326	0.01183	0.01052	0.00934	0.00828	0.00732
45	0.01771	0.01572	0.01391	0.01227	0.01079	0.00945	0.00826	0.00720	0.00626	0.00543
50	0.01551	0.01357	0.01182	0.01026	0.00887	0.00763	0.00655	0.00560	0.00478	0.00406
55	0.01373	0.01183	0.01014	0.00865	0.00735	0.00621	0.00523	0.00439	0.00367	0.00305
60	0.01224	0.01039	0.00877	0.00735	0.00613	0.00509	0.00420	0.00345	0.00283	0.00231

(Contd.)

144

Table D (*Contd.*)

Interest % (= 100r)

n (years)	6	6.5	7	7.5	8	9	10	12	15	20
1	1.00000	1.00000	1.00000	1.00000	1.00000	1.00000	1.00000	1.00000	1.00000	1.00000
2	0.48544	0.48426	0.48309	0.48193	0.48077	0.47847	0.47619	0.47170	0.46512	0.45455
3	0.31411	0.31258	0.31105	0.30954	0.30803	0.30505	0.30211	0.29635	0.28798	0.27473
4	0.22859	0.22690	0.22523	0.22357	0.22192	0.21867	0.21547	0.20923	0.20027	0.18629
5	0.17740	0.17563	0.17389	0.17216	0.17046	0.16709	0.16380	0.15741	0.14832	0.13438
6	0.14336	0.14157	0.13980	0.13804	0.13632	0.13292	0.12961	0.12323	0.11424	0.10071
7	0.11914	0.11733	0.11555	0.11380	0.11207	0.10869	0.10541	0.09912	0.09036	0.07742
8	0.10104	0.09924	0.09747	0.09573	0.09401	0.09067	0.08744	0.08130	0.07285	0.06061
9	0.08702	0.08524	0.08349	0.08177	0.08008	0.07680	0.07364	0.06768	0.05957	0.04808
10	0.07587	0.07410	0.07238	0.07069	0.06903	0.06582	0.06275	0.05698	0.04925	0.03852
11	0.06679	0.06506	0.06336	0.06170	0.06008	0.05695	0.05396	0.04842	0.04107	0.03110
12	0.05928	0.05757	0.05590	0.05428	0.05270	0.04965	0.04676	0.04144	0.03448	0.02526
13	0.05296	0.05128	0.04965	0.04806	0.04652	0.04357	0.04078	0.03568	0.02911	0.02062
14	0.04758	0.04594	0.04434	0.04280	0.04130	0.03843	0.03575	0.03087	0.02469	0.01689
15	0.04296	0.04135	0.03979	0.03829	0.03683	0.03406	0.03147	0.02682	0.02102	0.01388
16	0.03895	0.03738	0.03586	0.03439	0.03298	0.03030	0.02782	0.02339	0.01795	0.01144
17	0.03544	0.03391	0.03243	0.03100	0.02963	0.02705	0.02466	0.02046	0.01537	0.00944
18	0.03236	0.03085	0.02941	0.02803	0.02670	0.02421	0.02193	0.01794	0.01319	0.00781
19	0.02962	0.02816	0.02675	0.02541	0.02413	0.02173	0.01955	0.01576	0.01134	0.00646
20	0.02718	0.02576	0.02439	0.02309	0.02185	0.01955	0.01746	0.01388	0.00976	0.00536
25	0.01823	0.01698	0.01581	0.01471	0.01368	0.01181	0.01017	0.00750	0.00470	0.00212
30	0.01265	0.01158	0.01059	0.00967	0.00883	0.00734	0.00608	0.00414	0.00230	8.46-4*
35	0.00897	0.00806	0.00723	0.00648	0.00580	0.00464	0.00369	0.00232	0.00113	3.39-4*
40	0.00646	0.00569	0.00501	0.00440	0.00386	0.00296	0.00226	0.00130	5.62-4*	1.36-4*
45	0.00470	0.00406	0.00350	0.00301	0.00259	0.00190	0.00139	7.36-4*	2.79-4*	5.47-5*
50	0.00344	0.00291	0.00246	0.00207	0.00174	0.00123	8.59-4*	4.17-4*	1.39-4*	2.20-5*
55	0.00253	0.00210	0.00174	0.00143	0.00118					
60	0.00187	0.00152	0.00123	0.00010	0.00080					

*The figures -4 or -5 indicate that the figures preceding them should be multiplied by 10^{-4} or 10^{-5}.

Recommended Reading

Micro Economics

Samuelson, P. A. *Economics, an Introductory Analysis*, McGraw-Hill, New York (1976).
Nevin, E. *An Introduction to Micro Economics*, Croom Helm, London (1973).

Macro Economics

Samuelson, P.A. *Economics, an Introductory Analysis*, McGraw-Hill, New York (1976).

Cost Analysis

Introduction to Engineering Economics, Institution of Civil Engineers, London (1978).
Adler, H. A. *Economic Appraisal of Transport Projects*, Indiana University press, Indiana (1971).
Grant, E. L. and Grant Ireson E. *Principles of Engineering Economy*, Ronald Press, New York (1976).
Anon. *A Guide to the Economic Appraisal of Projects in Developing Countries*, HMSO, London, (1977).
Squire, L. & v.d. Tak, H. G. *Economic Analysis of Projects*, World Bank Research Publication. Johns Hopkins University Press, Baltimore, USA (1975).

Solutions to Problems

Problem P1.1

		$
(a)	*Annual fixed costs*	
	Annual repayments, using Table C = $4000/6.1446	650
	Garage, tax, etc.	250
		900
(b)	*Annual variable costs*	
	Fuel etc. 10 000 km at 3 cents/km	300
(c)	*Total annual costs*	
	FC($900) + VC($300)	1200

Therefore AC = $1200/10 000 = 12 cents/km

Problem P1.2

		$
(a)	*Annual fixed costs*	
	Annual repayments, using Table C = $4000/3.7908	1050
	Garage, tax, etc.	250
		1300
(b)	*Annual variable costs*	
	Fuel, etc. 20 000 km at 3 cents/km	600
(c)	*Total annual costs*	
	FC($1300) + VC($600)	1900

Therefore AC = $1900/20 000 = 9.5 cents/km

A comparison of the answers of Problems P1.1 and P1.2 shows that a car allowance, based on distance, is inequitable if different drivers travel different distances each year; whilst variable costs are the same, fixed costs are markedly different due to interest payment on capital investment.

Problem P1.3

The motorist will now drive only 10 000 km per year (Problem P1.1) instead of 20 000 (Problem P1.2) so his average costs will increase from 9.5 to 12 cents/km.

Considering a year's costs

		$
(a)	*Before using public transport*	
	20 000 km at 9.5 cents/km	1900
	or 9.5 cents/km	——
(b)	*After using public transport*	
	10 000 km by car at 12 cents/km (from Problem P1.1)	1200
	10 000 km by public transport at 4 cents/km	400
		1600

or 8 cents/km

Therefore annual saving = $1900 − 1600 $300

Thus, although public transport, at 4 cents/km, costs more than the motorist's variable costs for fuel etc. (3 cents/km), it pays him to use it. This is due to the saving in fixed costs.

Problem P1.4

The driver now buys his car for $8000, and sells it for $4000. Whether he borrows the money, or has it available but loses the interest, the problem is the same; he has to amortize $4000 over 10 years at 10% interest, but only has to borrow and repay the second $4000, as this will be available in cash when he sells the car. Thus, on the second $4000 his annual cost is merely the interest due, that is $400 p.a.

But he travels 10 000 km a year. Therefore the extra cost, over the cheaper car in Problem P1.1 is $400/10 000 = 4 cents/km.

Therefore the new average cost is $12 + 4 = 16$ cents/km.

Problem P2.1

(a) Firstly Table P2.1 should be compiled.
(b) The second step is to construct Figure P2.1 with the AC curve.
(c) The third step is to complete Table P2.1(b), in increments of 1 m^3, to determine MC. The AC curve has been rounded off, and output costs are read off the AC curve.
(d) The MC curve can now be plotted on Figure P2.1. It can be seen that minimum costs of $\$22.7/\text{m}^3$ can be achieved with 1 mixer and 4 labourers, at a daily output of 14 m^3.

Table P2.1(a) Output of a mixer

Units of labour	Daily output (m³)	Daily costs ($)			Total costs, TC ($)	Average costs, AC ($/m³)
		Labour	Plant	Materials		
1	2	40	20	20	80	40
2	5.5	80	20	55	155	28.2
3	9.5	120	20	95	235	24.7
4	14	160	20	140	320	22.9
5	16.5	200	20	165	385	23.3
6	19	240	20	190	450	23.7
7	21	280	20	210	510	24.3
8	23	320	20	230	570	24.8

Table P2.1(b) Calculation of marginal costs

Daily output (m³)	Average cost, AC ($)	Total cost, TC ($)	Marginal cost, MC ($)
11	23.5	258.5	
			18.7
12	23.1	277.2	
			19.2
13	22.8	296.4	
			21.4
14	22.7	317.8	
			24.2
15	22.8	342.0	
			26.0
16	23.0	368.0	

The contractor has to produce 37.5 m³ per day. If he uses 1 mixer, costs will be infinite; indeed, 1 mixer will probably not suffice. If he uses 2 mixers, it can be seen from Figure P2.1 that average costs would rise to $23.6/m³, and he would need 12 labourers.

If he uses 3 mixers, with 13 labourers, he would achieve minimum costs with 2 mixers, each with 4 labourers, and 1 mixer with 3. Outputs would be 28 m³ at $22.7 and 9.5 m³ at $24.8, the average cost being

$$\$\frac{28 \times 22.7 + 9.5 \times 24.7}{37.5} = \$23.2/m^3$$

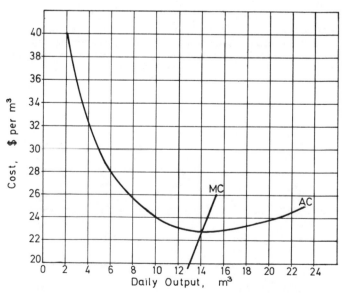

Figure P2.1   AC and MC curves for a concrete mixer

Problem P2.2

The first step is to calculate and compile Table P2.2 from hourly costs and output data. The marginal cost is the same irrespective of plant hire costs.

These data are then plotted in Figure P2.2.

At $50/hour minimum costs are achieved at point A, the nearest whole number of men

Table P2.2 Hourly costs

No. of men	Cost of labour	Plant cost ($)		Total cost ($)		Output (m²)	Average cost ($/m²)		Marginal cost ($/m²)
							Plant at 50 $/ hour	Plant at 100 $/ hour	
2	10	50	100	60	110	60	1.0	1.83	
									0.08
3	15	50	100	65	115	125	0.52	0.92	
									0.09
4	20	50	100	70	120	180	0.39	0.67	
									0.25
5	25	50	100	75	125	200	0.38	0.63	
									0.42
6	30	50	100	80	130	212	0.38	0.61	
									0.63
7	35	50	100	85	135	220	0.39	0.61	
									1.00
8	40	50	100	90	140	225	0.40	0.62	

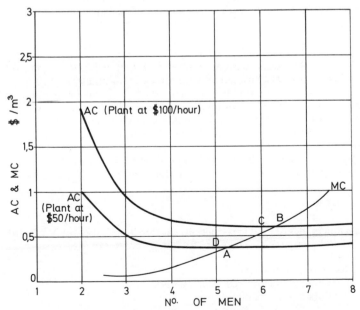

Figure P2.2   Hourly costs of a surfacing contractor

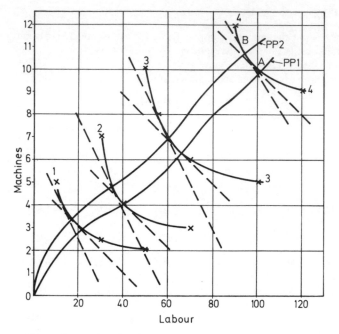

Figure P3.1   Production paths for a pump manufacturer

being 5 at point D with costs of $0.38/m², whilst minimum costs with plant hire at $100/hour will be $0.61/m² with 6 men in attendance, at point C being nearest to B.

Problem P3.1

On Figure P3.1 the output data are plotted and contour lines (isoquants) are sketched between them. Parallel isocost lines are then drawn tangentially to the isoquants at slopes shown, such that equal expenditure is incurred at any point thereon. For example, at a machinery cost of $200/day, with $1000 available, the isocost line will pass through 5 machines, and zero labour, or on 50 labourers and zero machines. With the halving of machine hire to $100/day, it will pass through 10 machines and no labour, but still through 50 labourers, and no machines.

The production paths, PP1 and PP2, are sketched through the intersection of parallel isocost and of isoquant lines: for daily machine costs of $200 and $100 respectively.

Problem P3.2

With machinery at $200/day, production costs are lowest at point A, with 10 machines at $200, and 100 men at $20. Therefore total daily costs are

	$
$200 \times 10 + 20×100	4000
Therefore production costs per pump, excluding materials, etc.	1000

With machinery at $100/day, minimum costs result at B, but, since only whole numbers of machines can be used, 11 machines would be hired with 92 men. Therefore total daily costs are

	$
$100 \times 11 + 20×92	2940
Therefore cost per pump	735

Problem P3.3

Cost of 4 pumps a day

	$
$100 \times 11 + $20 \times 92 + $100 \times 4 + 300	$3640

Cost of 3 pumps a day

$100 \times 7 + $20 \times 60 + $100 \times 3 + 300	2500

Cost of 2 pumps a day

$100 \times 5 + $20 \times 35 + $100 \times 2 + 300	1700

Income from sales

4 pumps a day	4000
3 pumps a day	3300
2 pumps a day	2400

Profits from sales

Daily profits are the excess of sales price over cost.

At 4 pumps a day, profit $= $4000 - $3640 = 360
At 3 pumps a day, profit $= $3300 - $2500 = 800
At 2 pumps a day, profit $= $2400 - $1700 = 700

The manufacturer would maximize his profit at $800 a day by producing 3 pumps a day with 7 machines and 60 men. It is noted that his profit would reduce by increasing production due both to the reduced sale price of each pump, and to the law of diminishing returns.

Problem P4.1

The industry will reach a balance when the S and D curves cross and both formulae apply. Thus

$$\$P = \frac{Q}{1000} = \frac{1000}{Q}$$

and so

$$Q = \$1000 \quad \text{and} \quad P = \$1$$

Problem P4.2

The supply curve $\$P = \$Q/1000$ will remain unaltered, but, at any given price $\$P$, double the number of cups will be required. Thus the new demand curve will be required. Thus the new demand curve will be $\$P = \$2000/Q$.

At the new intersection

$$\$P = \frac{\$2000}{Q} = \frac{\$Q}{1000}$$

Thus

$$Q = \$1410 \quad \text{and} \quad P = \$1.41$$

Note that both price and quantity increase, but not to double.

Problem P5.1

The data should be plotted, as in Figure P5.1, and the D and S_1 curves derived. The intersection indicates that 620 litres would be sold daily at a price of $1.4/litre. With rezoning, transport costs raise the supply curve from S_1 to S_2, S_2 lying at 50 cents above S_1. The new market is for a reduced quantity of 560 litres at the increase of $1.7/litre, the full increase of 50 cents per litre not being paid, as demand is not infinitely inelastic.

Problem P5.2

Since there is perfect competition there is, by definition, a large number of firms, and the moving of one firm would not affect either the price, or the quantity sold.

The market price, as can be seen from Figure P5.1 will remain at $1.4/litre, but the individual firms supply curve will shift to S_2 with a minimum supply price of $1.5/litre. The firm will be unable to compete and will leave the market.

Problem P6.1

(a) Firstly the marginal cost, MC, should be calculated as follows for the industry in Table P6.1.

(b) These data are plotted in Figure P6.1 for the industry. The MC curve is the industry's

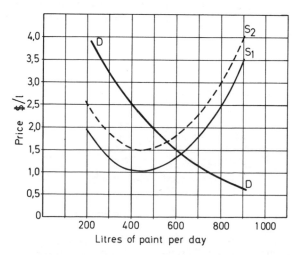

Figure P5.1   Effect of rezoning of the paint industry

Table P6.1

No. of bicycles per day produced in industry	Average cost, AC ($)	Total cost, TC ($)	Marginal cost, MC ($)
100	53	5 300	
			39
200	46	9 200	
			34
300	42	12 600	
			34
400	40	16 000	
			50
500	42	21 000	
			78
600	48	28 800	
			111
700	57	39 900	

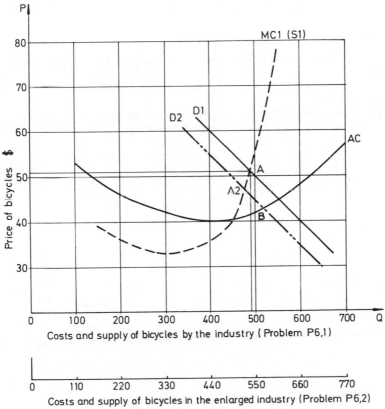

Costs and supply of bicycles by the industry (Problem P6,1)

Costs and supply of bicycles in the enlarged industry (Problem P6,2)

Figure P6.1

supply curve, and, as there are 10 similar firms in the market, if the x-axis (Q) scale is divided by 10, the MC curve becomes the firm's supply curve.

(c)   When the market demand curve, DD, is plotted the intersection (at A) gives the market as 490 bicycles a day (49 from each firm) at $51.

(4)   Whilst the sales price is $51 ($AC$) the cost price at B is $41, giving a profit per bicycle of $10, or, per firm, of $490 a day.

Problem P6.2

In this case the output of the industry has increased 10% but costs of each factory are unaltered: for any given cost, be it marginal or average, production will have increased 10%. The plot shown in Figure P6.1 can be used with a new x-axis, as shown. However, the demand curve has remained unaltered, so it must be replotted as D_2 to the new scale. The new intersection of the MC (supply) curve is now at A_2: a production of some 520 bicycles a day at a price of $47 each, the corresponding average cost being slightly less than before at about $40.5.

Thus the industry's profit will become 520($47 − $40.5) = $3380. Each firm will now make and sell 520/11, say 47, bicycles at a daily profit of $305

Thus the effect of the extra firm entering the market will be to increase the supply from 490 to 520 a day, to cut each firms profit from $490 to $305 daily. It should be noted that this is not truly perfect competition, as the new firm has affected the market.

Problem P6.3

(a)   Since all firms are similar, Table P6.3 can be compiled. Since MC is the supply curve, the industry's supply curve can be drawn of Figure P6.3.

(b)   Since the demand is of constant elasticity a straight line demand will exist. This can thus be drawn.

(c)   This crosses the supply curve at $175 with a daily demand of 70 motor scooters.

(d)   Thus each of the equal 10 firms can produce and sell 7 motor scooters a day at a price of $175.

Table P6.3

Daily output of industry Q	Average costs ($)	Total costs ($)	Marginal cost ($)
20	180	3 600	
			120
30	160	4 800	
			100
40	145	5 800	
			70
50	130	6 500	
			100
60	125	7 500	
			160
70	130	9 100	
			210
80	140	11 200	

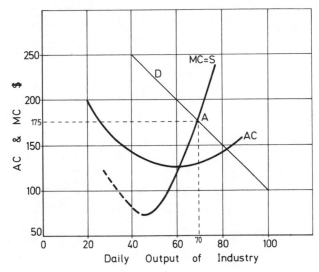

Figure P6.3   *AC* and *MC* curves of industry

Problem P7.1

(a) Complete Table P7.1.
(b) Plot Figure P7.1.
(c) Profit will be a maximum where LRMC crosses the MR curve, i.e. at 7 Ml/day and a price of 12.8 cents/l.
(d) Total daily revenue will be 12.8 cents/l × 7 Ml = $895 000
and total costs 8.0 cents/l × 7 Ml   = $560 000
Therefore the maximum daily profit will be $335 000 when output is restricted to 7 Ml/day.

Table P7.1

Output (Ml/day)	LRAC (cents/l)	TC (10^4 $)	LRMC (cents/l)
3	9.0	27	
			7.5
5	8.4	42	
			7.0
7	8.0	56	
			8.0
9	8.0	72	
			12.0
11	8.7	96	

Problem P7.2

The situation is illustrated in Figure P7.2. The monopolistic situation is not normal, in that price is fixed at a level below that at which profit could be maximized, since, at this

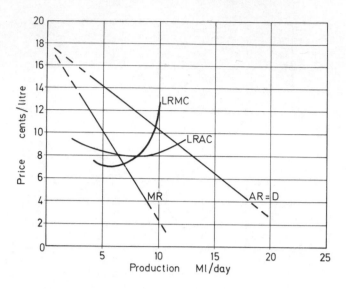

Figure P7.1   Maximization of profits of an oil cartel

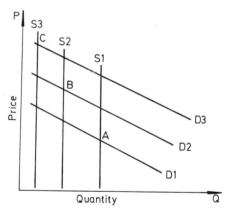

Figure P7.2   Determination of oil price

price, inflation would become so severe that the income which OPEC countries would accumulate and invest overseas would inflate and thus devalue too severely.

In the figure D_1 represents, say, the 1980 curve. With increasing users and demand it will increase to D_2 by, say, 1990, and D_3 by, say, 2000.

Similarly, S_1 represents the 1980 supply curve. The price could reach A, but, for reasons above, has been fixed below this level. By 1990 S_2 will result as a supply curve, and by 2000, S_3. The quantity available for annual sale will reduce as the known reserves dry up and a price of C, not one *fixed* under C, may well result as, by then, the sellers will be forced to maximize their profit on their dwindling reserves. The price and supply position will move in a general direction from below A to C.

Problem P8.1

(a) When the economy is contracting, few entrepreneurs are willing to construct new buildings. Since most architects (and structural engineers) are concerned mainly with new works, their profession is extremely cyclic. Civil engineering schemes are also likely to be postponed, but, as such works are often long-term projects, the employment is less cyclic than architecture. Moreover, much civil engineering is concerned with maintenance, most of which must proceed as an insurance on the original capital investment. Mechanical engineering is less cyclic, as much employment is concerned with the production of essential consumable items, such as food, and again, much is also concerned with maintenance.

Electrical engineering is even less cyclic as demands for power slacken little during recessions; however, the demand for electronic goods may well be cyclic, so, on average, there may be little to choose between electrical and mechanical engineering.

(b) Governments normally maintain their cadre of civil engineers to plan, and to cope with essential works at the troughs of cycles. When boom conditions return they are insufficient in number or expertise to handle major projects, and governments then engage consultants. Thus consulting engineering is extremely cyclic in nature. However, where consultants are engaged on major projects which take many years for analysis and design, their services may be retained through depressions.

Contracting is also cyclic, but not to the extreme of consultancy, due to the need for maintenance, even during depressions.

Government staff are seldom discharged during recessions — to do so would add to unemployment and make recruitment on upturns difficult — and even government employees are retained where possible. This staff will undertake essential maintenance, and such little construction work as is available. Such employment is only minimally cyclic.

Problem P12.1

(a) *Costs.* Since the deviation will last forever, the capital need not be repaid, but only the interest. Therefore annual costs are $5\% \times 1\,000\,000 = \$50\,000$ p.a.

(b) *Savings on train movements.* The financial saving, per train movement, is $\$10$, being $\$5$ in local, and $\$5$ in foreign currency, since the distance saving is 1 km. However, the *economic* saving involves the multiplication of the foreign currency saving by the shadow price, and is $\$5 + 3 \times \$5 = \$20$ per movement.

(c) *Calculation of train movements.* Let n be the minimum number of train movements per year for savings to equal costs. Therefore

$$n \times \$20 = \$50\,000$$

and so

$$n = 2500 \; movements \; per \; year.$$

Problem P12.2

If the railway were privately owned, the operators would only be concerned with the financial saving of $\$10$ per movement, not the economic saving, as they have no interest in foreign currency, only with profit maximization (with existing fares). Let n be the minimum number of annual movements to justify the deviation. Thus

$$n \times \$10 = \$50\,000 \; (annual \; cost)$$

and so

$$n = 5000$$

But only 3000 movements take place, so the shortening scheme would not be justified.

Problem P12.3

At 4000 trains per year, at a financial saving of $10 per train per km, the annual saving to the operator would be $4000 \times \$10 = \$40\,000$. But annual costs due as capital interest would be $50\,000. Thus the operator would need a subsidy of $10\,000 p.a., in local currency, from the state. In turn, the state would save $4000 \times \$5 = \$20\,000$ in foreign currency commitments, which would be worth $60\,000 in local currency to the state, since the shadow price is 3. Thus the state could well justify the necessary minimum subsidy of $10\,000, and could consider any subsidy up to $60\,000 p.a.

Problem P13.1

(a) *Initial costs.* The total costs are the $1\,000\,000 for development plus $100\,000 compensation, since government cannot be expected to compensate the adjacent landowners who would suffer. Similarly, the land value appreciation on land near the road will not be paid by the landowners to the mine developer. (Indeed, if the latter wishes to gain from this he should either buy the land before development, and subsequently sell it at a profit, or he should ask government to contribute towards the road cost.) Thus the necessary development capital is $1\,100\,000.

   The entrepreneur will borrow this capital at 6% p.a. (If he had the capital he would forego the interest gained by otherwise investing it in the market, which amounts to the same thing.) Since the scheme is long lasting he will not have to allow for capital redemption, but only pay the annual interest, being 6% of $1\,100\,000, i.e. $66\,000 p.a.

(b) *Annual cash flow*

		$
Expenditure		
Interest on capital		66 000
Operating costs		100 000
	Total	166 000
Income		
Sales		200 000
Profits		
Sales income less expenditure		34 000

Note: The annual profit of $34\,000 in a market where the borrowing rate is 6% p.a. is equivalent to $34\,000/0.06 or $567\,000 in cash.

Problem P14.1

(a) At the end of year 1, 20% of the equipment life will have been used, and the replacement value of new equipment will be $100\,000 \times 1.1 = \$110\,000$. Thus the then current value of wear will have been $20\% \times \$110\,000 = \$22\,000$. Since 10 000 bicycles have been made the equipment on cost is $2.2 per bicycle — the increase in sales price will be 20 cents.

(b) Similarly, at the end of the second year the replacement value will be $121\,000 of which 20% additional wear will have been carried out, being $24\,000 or $2.42 each, the yearly increase in sales price being 22 cents.

(c) By similar reasoning it can be calculated that the on-cost per bicycle produced will be $2.66 for the third year's production, $2.93 for the fourth, and $3.22 for the fifth, the yearly increases in sales price being 24, 27, and 29 cents respectively.

Problem P14.2

Consider the costs of each bicycle made in the second year.
(a) *Plant*. Replacement costs were determined from Problem P14.1 at $2.42.
(b) *Labour*. Labour during the second year will be $10 \times 1.1 = \$11.00$
(c) *Materials*. During the first six months of operation materials cost was $10, during the second 10×1.05, during the third $10 \times 1.05^2 = \$11.03$ and during the fourth $10 \times 1.05^3 = \$11.58$. Thus during the third and fourth six-monthly periods, the cost averaged $11.30.
(d) *Total costs*

	$
Plant	2.42
Labour	11.00
Materials	11.30
	24.72
Overheads etc. 50%	12.36
Sales price in second year	37.08

Problem P16.1

The dollar nominal value of the share is of little importance. The investors are able to buy or sell at $1.20 and earn 10 cents a year: a return of 8.33%. Had they invested in government stock they would have received 5%. In other words they expect $8.33/5 = 1.67$ times as much interest in the private sector, due to risk.

When the government stock increases its interest rates to 7.5%, investors, other things being equal, will still want 1.67 times this rate as a compensation for risk, that is 12.5%.

Thus, in the long term, the share price will drop from 120 to 80 cents, with the return of 10 cents.

Problem P16.2

Since the scheme is long lasting the capital need not be repaid, only the interest. Therefore the annual capital costs are 9% of $1 million, or $90 000.

Operating costs are $80 000 p.a., making total costs $170 000, and profits $80 000 p.a. or 8% of the investment. Were the interest rate to rise to 10%, the profit would still be $70 000, or 7% clear. Indeed, the project would not show a loss at all rates below 17% when annual interest payable ($170 000) plus operating costs ($80 000) would equal anticipated income ($250 000). This indicates that the scheme is not sensitive to interest rate changes.

Problem P16.3

(a) Using Table C we can calculate the fixed repayments of capital costs as $1 000 000/9.1285 = \$109 500$ p.a. This will apply in all years unless the market investment rate changes.
(b) *Annual surplus of income over operational costs*. Without inflation this would be

$200\ 000 - \$80\ 000 = \$120\ 000$ p.a. With inflation at 10% p.a., this becomes

$$\$120\ 000 \times \left(\frac{100 + 10}{100}\right)^n = \$120\ 000 \times 1.1^n$$

at the end of year n (during the nth year).

(c) Nett income after year n is

$$\$120\ 000 \times 1.1^n - \$109\ 500$$

(d) Nett income after year n, in values of year n, discounted to PDV, at a market rate of 9%, is

$$\$\frac{120\ 000 \times 1.1^n - 109\ 500}{(1.09)^n}$$

since any money n years hence be divided by $[(100 + r)/100]^n$ where r is the interest rate, to give it its present day equivalence.

(e) This income is in inflated dollars. To obtain the value in current dollars it must be divided by $[(100 + R)/100]^n$ where R is the inflation rate. Thus nett income in year n at PDV becomes

$$\$\frac{120\ 000 \times 1.1^n - 109\ 500}{1.09^n \times 1.1^n} = \$\frac{120\ 000 \times 1.1^n - 109\ 500}{1.2^n}$$

Problem P17.1

(a) *Scheme A: self-built.* Consider the PDV as being completion date, the end of the third year. During construction, the developer will not have paid back interest, and will have owed $10\ 000 for 3 years, $10\ 000 for 2 years, and $10\ 000 for 1 year. He thus owes $10\ 000(1.1^3 + 1.1^2 + 1.1) = \$36\ 400$. But, that day, the $4000 rent in advance will be payable, reducing debts to $32\ 400.

(b) *Scheme B contractor built.* After 1 year the building society will pay the builder $35\ 000, but $4000 rent will be received, reducing debts to $31\ 000. After 2 years the debt will be $31\ 000 \times 1.1$ less $4000 = \$30\ 100$, and after the third year $30\ 100 \times 1.1$ less $4000 = \$29.100$.

(c) *Comparison of schemes.* It can be seen that after the third year, when equal rental income accrues from both schemes, the contractor-built scheme, although costing $5000 more than the owner-built scheme, will result in $3300 less debt, i.e. greater profitability.

Problem P17.2

Take the end of the third year, that is the beginning of the fourth, as the present day.

(a) *Scheme A: self-built.* Amount owing $32\ 400. Repayments (now at year end) $4000 p.a. The capital/repayment ratio is 8.1. From Table C, using 10% interest, it can be seen that this corresponds to approximately $17\frac{1}{2}$ years at $4000 (year-end) payments to redeem the PDV of $32\ 400.

(b) *Scheme B: contractor built.* Amount owing $29\ 100. The capital/repayment ratio of 7.28, which from Table C, corresponds to approximately $13\frac{1}{2}$ years.

(c) *Comparison.* It can be seen that the loan for the more expensive contractor-built scheme can be repaid, with interest, 4 years sooner.

Problem P17.3

(a) Calculate the PDV of the replacement value after 5 years. Using Table B, this can be calculated as $5000 \times 0.78353 = $3920 at 5% and as $5000 \times 0.62092 = $3100 at 10% interest rates.
(b) Thus the purchase of the $10 000 machine is the same as the purchase of a $6080 machine at 5% or of a $6900 machine at 10% interest.
(c) The PDV of 5 year-end receipts of income after 5 years, can be determined, using Table C, as $2000 \times 4.3295 = $8660 at 5% or $2000 \times 3.7908 = 7580 at 10%.
(d) Thus the B/C ratios are 8660/6080 = 1.42, and 7580/6900 = 1.10 at 5% and 10% interest respectively.

Problem P18.1

(a) Although the infrastructure is an everlasting investment, it will be valueless *to the entrepreneur* once the scheme has reached the end of its life in 15 years. Thus all the capital investment, $10 million, must be written off in 15 years.
(b) For simplicity take the PDV as the commencement of the project. The annual operational and sales income must be discounted forward for 15 years. Using Table C the factors are 7.6061 for a 10%, and 6.8109 for a 12% interest rate.
(c) Table P18.1 can now be compiled, in $ millions at PDVs.
(d) Although it is not very sensitive to interest rate changes, the project would not be viable if it were a mine, as the B/C ratio is too low.

   As an agricultural development, since food prices are not liable to change, and forecasting is more reliable, the project would be viable. (Agricultural schemes would, in practice, only have short lives if the land were planned for more profitable use at a later date.)

Problem P18.2

If inflation occurred both the annual operational expenses, and annual sales would increase in monetary value. For example, at 5% inflation, the first year's nett income would increase from $2 million to $2.10 million, and the second year's to $2.205 million, etc., whilst the PDV would increase from $1.79 to $1.88 million for the first year and from $1.59 to $1.78 million for the second year.

   However, in inflated money, these enhanced values are worth less at today's spending power, so, from this aspect, there would be no change in viability.

   As far as the investment is concerned it could be argued, assuming straight line depreciation, that without inflation the investment after 1 year would be worth 14/15 or $9.33 million, but with 5% inflation $9.80 millions, i.e. that only $0.2 million depreciation

Table P18.1

Interest	10%			12%		
Investment	Op. costs	Total costs	Sales	Op. costs	Total costs	Sales
10	7.61	17.61	22.82	6.81	16.81	20.43
B/C ratio	$\frac{22.82}{17.61} = 1.30$			$\frac{20.43}{16.81} = 1.22$		

had occurred. This is fallacious as, according to current cost accounting, historic values must be forgotten at the year end, when depreciation would be taken as 1/15th of the new value of $10.5 million, being $0.70 million, which is in inflated money, being worth only $0.67 million in spending power of the original money — which was the original estimate of depreciation.

On balance, then, the inflation has not affected the viability or profitability of the scheme.

Problem P18.3

(a) *Capital costs.* Life is indefinite and thus only the annual interest on the loan need be considered, since the capital never need be repaid.

(b) *Private enterprise.* The entrepreneur cannot gain from other users of the infrastructure, and is responsible for all loan interest repayment, being 10% of $10 million, i.e. $1 million p.a. His total annual costs also include operational costs of $1.2 million, and are $2.2 million. The anticipated sales are $2.5 million p.a. and his B/C ratio of $2.5 million/2.2 million $= 1.14$ which is not sufficiently attractive to make the scheme viable.

(c) *Public enterprise.* Similarly, it can be shown that the public B/C ratio is

$$\frac{\$2.5 \text{ million}}{\$1.2 \text{ million} + 0.6 \text{ million}} = 1.39$$

which makes the scheme more attractive.

In addition, the infrastructure will be developed, which will add to other land values and increase sales or profits from adjacent land. In turn taxation income to government will increase. These are externalities. In this problem they cannot be quantified, as no data are given, but it can be concluded that the B/C ratio *to the state* will exceed 1.39, so the scheme become quite attractive.

Problem P19.1

Annual costs remain the same at $1.8 million, but benefits will increase by the extra taxation of $500 000 p.a. generated on landowners extra income of $1 million p.a., to $3 million. The B/C ratio now becomes 1.67, making the scheme very attractive.

Problem P19.2

(a) *Annual financial costs.* These will increase from $1.8 million to $1.9 million, due to the transport for export, but the transport is paid for in foreign currency with a shadow price of 2, so the total *economic* costs will be $2 million.

(b) *Annual financial income.* This will remain at $3 million, but includes $1.25 million in export sales which must be multiplied by the shadow price of 2 to make the *economic* income $4.25 million.

(c) *The economic B/C ratio.* Thus becomes $4.25/2 = 2.12$.

Problem P19.3

(a) Complete Table P19.1

(b) *B/C ratios.* It is necessary to consider the benefits of schemes B and C, both of which involve costs, but the benefits are less tangible. However, the road users under scheme A are prepared to spend $2.5 million p.a. using the existing road, so it is

Table P19.1 Annual road costs ($1000)

Scheme	A	B	C
Interest on construction	—	500	900
Purchase of land	—	250	200
Sale of land	—	100	100
Vehicle and social costs	2500	1500	1000

legitimate to regard this as a benefit. The other benefit is the sale of land on scheme A, the existing road.

The costs can be taken as the sum of construction costs, land acquisition costs, and the new user costs. Thus the B/C ratios might be calculated as follows:

$$\text{Scheme B} \quad \text{B/C ratio} = \frac{\text{total benefits}}{\text{total costs}} = \frac{2500 + 100}{500 + 250 + 1500} = 1.16$$

$$\text{Scheme C} \quad \text{B/C ratio} = \frac{2500 + 100}{900 + 200 + 1000} = 1.24$$

(c) *Differential B/C ratios.* These constitute the normal approach to road justification schemes and are the change in road user savings divided by the change in road costs:

$$\text{Scheme B} \quad \text{Differential B/C ratio} = \frac{2500 - 1500}{500 + 250 - 100} = 1.54$$

$$\text{Scheme C} \quad \text{Differential B/C ratio} = \frac{2500 - 1000}{900 + 200 - 100} = 1.50$$

(d) *Analysis.* Both schemes are favourable, and could be adopted. Scheme B has a higher differential B/C ratio, and might be preferred, but the difference (1.54 compared with 1.50) is only marginal, and within the limits of forecasting error.

Scheme C, on the other hand, has a marginally higher B/C ratio (1.24 as compared with 1.16) and thus deserves a marginally higher priority. Thus either scheme would be acceptable, and the choice could well be based on engineering or political, rather than on economic, grounds.

Problem P20.1

(a) *Financial.* In this case we are concerned only with financial matters. Consider the 15m^2 occupied by a car.

Annual costs per car (15m^2)

	$
Road construction $15 \times \$10 \times 5\%$	7.5
Road maintenance $15 \times \$0.5$	7.5
Meter hire	25
Supervisor $\$5000/250$	20
	60

Annual income per car

1500×5 cents	$75
B/C ratio $= 75/60$	1.25

(b) *Economic.* The land value can be considered as the social cost. Land values are high, to the direct benefit of local landowners etc., solely due to the convenience of the transport system, including the parking. The land occupied by the parked car belongs to the citizen, not the local shopkeeper, nor the motorist. In allowing the motorist to park, the average citizen is forfeiting the use of the public land — its use can be regarded as a social diseconomy, if the B/C ratio, including land value, is less than unity.

The annual land use, per car, costs $5\% \times 15 \times 100 = \$75$. Therefore the *economic* B/C ratio is

$$\$\frac{75}{75 + 60} = 0.56$$

(c) *Analysis.* To incur no social diseconomy, total costs should equal total benefits. Total costs are $135 p.a., and parking is for 1500 hours p.a. Therefore the parking fee should be $135/1500 = 9.0$ cents/hour. The hidden subsidy is thus $1500(9.0 - 5)$ cents or $60.00 per year per car. This could, of course, be recovered by raising rates on property near the parking facility when this is implemented.

Problem P20.2

(a) *The subsidy.* The financial B/C ratio must be 1 (or more) after the subsidy is paid, i.e. costs must not exceed sales plus subsidy.

Since the scheme lasts 40 years, this approximates to everlasting, and only interest need be paid, i.e. annual cost analysis is adequate.

Annual costs	$ million
10% of $50 million	5.0
5% of $60 million	3.0
Local operational costs	1.0
Expatriate costs	1.0
	10.0
Less: annual sales	8.0
Subsidy (loss)	2.0

Thus the state industry would show a loss, i.e. need subsidizing, at $2 million p.a..

(b) *Shadow pricing.* Let P be the shadow price. Then annual economic costs (in $ millions) are

	$ millions
10% of $50 million $\times P$	5P
5% of $60 million	3
Local operational costs	1
Expatriate costs	1P
	4 + 6P

and annual economic income is $8P$ million. But for marginal viability, costs = income and so

$$4 + 6P = 8P \quad \text{i.e.} \quad P = 2$$

Thus the scheme is economically viable if the shadow price is 2, and there would be a nett gain of $8 - 6$ million $= \$2$ million in foreign currency for a subsidy of $2 million in local currency each year.

Problem P21.1

For convenience, consider the day of completion of development, when sales start, as the PDV.

		$ millions
(a)	*Costs*	
	Land purchase $1 000 000 1 year at 10%	1.10
	First expenditure $500 000, $\frac{1}{2}$ year at 10%	0.52
	Final expenditure	0.50
	Operating expenses (use Table B)	
	1st year $300 000, discounted at 10% for 1 year	0.27
	2nd year $300 000, discounted at 10% for 1 year	0.25
	3rd year $300 000, discounted at 10% for 1 year	0.22
	4th year $300 000, discounted at 10% for 1 year	0.20
		———
	Total PDV expenditure	3.06
		———

(b)	*Sales*	
	Income from 1st year's sales	
	$\quad$ 500 000 × $3, discounted at 10%	1.36
	Income from 2nd year	
	$\quad$ 500 000 × $3 × 1.15, discounted at 10%	1.43
	Income from 3rd year	
	$\quad$ 500 000 × $3 × 1.15^2 discounted at 10%	1.49
	Income from 4th year	
	$\quad$ 500 000 × $3 × 1.15^3 discounted at 10% 1.56	1.56
	Land sold after 4 years	
	$\quad$ 100 000, discounted at 10%	0.07
		———
	Total PDV income	5.91
		———

(c) *B/C ratio.* This is income/costs at any PDV, in this case the PDV on completion $= 5.91/3.06 = 1.93$.

Problem P21.2

Take the commissioning date as the present day.

		$ millions
(a)	*Costs (PDV)*	
	Plant	1.000
	Labour and materials	
	$\quad$ Year 1 : $10 × 8000 × 0.926	0.074
	$\quad$ Year 2 : $10 × 7000 × 0.926^2	0.060
	$\quad$ Year 3 : $10 × 6000 × 0.926^3	0.048
	$\quad$ Year 4 : $10 × 5000 × 0.926^4	0.037
	$\quad$ Year 5 : $10 × 4000 × 0.926^5	0.027
		———
	Total	1.246
		———

(b)   *Income*
    Sales of castings
        Year 1: $80 × 8000 × 0.926   0.593
        Year 2: $70 × 7000 × 0.926^2 0.420
        Year 3: $60 × 6000 × 0.926^3 0.286
        Year 4: $55 × 5000 × 0.926^4 0.202
        Year 5: $50 × 4000 × 0.926^5 0.136
    Sale of plant
        $200 000 × 0.926^5 0.136
                                                                        _____
    Total income   1.773
                                                                        _____

(c)   *B/C ratio.* This is $1.773/1.246 = 1.42$. The PDV of profits is $530 000.

Problem P21.3

With the onset of inflation, it is likely that not only sales, but labour and materials will similarly inflate. If the market interest role were to similarly increase, all costs and benefits, and hence profits, would increase pro rata; the B/C ratio would remain unaltered, and the profitability unchanged.

   However, market interest rates seldom rise proportionately with inflation—if they did we would have no trouble in safeguarding pensions. Thus with no, or lesser rises in market rates, the percentage profit, and hence the B/C ratio will increase, as costs will not rise proportionately. Against this, if the plant has to be replaced after 5 years, the replacement will cost more, and some of the extra profits should be set aside to offset this future increase. Thus the B/C ratio would remain largely unchanged.

Problem P22.1

We are not concerned with costs *per se* (the B/C ratio is unity), only with extra costs of building a second runway to achieve 2% more service ability.
(a)   *Annual costs saved by a single runway layout.* The extra annual cost of a second runway to last 10 years, at 5% interest rate, can be determined from Table C as

$$\$\frac{10\ 000\ 000}{7.7217} = \$1\ 295\ 000.$$

(b)   *Annual extra costs incurred by a single runway layout.* Since the undertaking is state owned, all costs must be considered, including social costs, even though the state may not pay compensation. Let n be the yearly movements. The total cost per diversion is $4000. An extra 2% of movements will be diverted, thus total annual cost of diversions is $4000 × n/50 = $80n$.
(c)   *Break-even point.* At this point the B/C ratio is unity. Thus the benefits (savings) = costs. Therefore $1 295 000 = $80n$, and so $n = 16\ 187$ movements, i.e. 44 movements per day.

   It can thus be concluded that with less than 40 daily movements the engineer should recommend a single runway, and with more than 50 a double runway layout. The choice of design between 40 and 50 daily movements would be marginal in view of the lack of precision associated with the quantification of social costs.

Problem P22.2

(a)   Three assumptions must be made: firstly that running costs of a standby plant are so small that they can be neglected; secondly that its usage will be so little that its life will be infinite; and thirdly the number of working days per year — assume this to be 250.

(b) *Cost of standby plant.* Let the initial cost be P. Then annual costs will be $0.06P$.

(c) *Cost of shutdown.* This is the same as the gain if no shutdown resulted.

There are 250 working days per year at a 0.1% risk of stoppage. Thus on average there will be 0.25 days lost per year involving a loss of $4000. It follows that annual losses without a standby plant cost $1000.

(d) *Analysis.* It follows that, for justification, $0.06P = \$1000$, i.e. $P = \$16\ 700$

Thus the plant at $15 000 should be purchased.

Problem P22.3

(a) *Capital costs.* The dyke would be everlasting, so only annual costs need be considered.

(b) *For 10% interest rate.* We can now compile Table P22.3. In this case the objective must be to minimize annual costs, so at 10% interest rate the design height should be 6 metres.

(c) *For a 20% interest rate.* A similar table could be deduced, from which it could be determined that the design height should be reduced to 4.5 metres.

Table P22.3   Costs of a dyke

Height of dyke (m)	2	3	4	5	6	7
Annual interest ($1000)	1.0	2.5	4.0	6.5	10.0	15.0
Annual cost of floods ($1000)	30	20	10	5.0	1	0.5
Total annual costs ($1000)	31	22.5	14.0	11.5	11.0	15.5

Index